Compendium of Plant Genomes

Series Editor

Chittaranjan Kole, Raja Ramanna Fellow, Government of India, ICAR-National Research Center on Plant Biotechnology, Pusa, New Delhi, India

Whole-genome sequencing is at the cutting edge of life sciences in the new millennium. Since the first genome sequencing of the model plant Arabidopsis thaliana in 2000, whole genomes of about 100 plant species have been sequenced and genome sequences of several other plants are in the pipeline. Research publications on these genome initiatives are scattered on dedicated web sites and in journals with all too brief descriptions. The individual volumes elucidate the background history of the national and international genome initiatives; public and private partners involved; strategies and genomic resources and tools utilized; enumeration on the sequences and their assembly; repetitive sequences; gene annotation and genome duplication. In addition, synteny with other sequences, comparison of gene families and most importantly potential of the genome sequence information for gene pool characterization and genetic improvement of crop plants are described.

Interested in editing a volume on a crop or model plant?
Please contact Prof. C. Kole, Series Editor, at ckoleorg@gmail.com

More information about this series at http://www.springer.com/series/11805

Shengyi Liu • Rod Snowdon •
Chittaranjan Kole
Editors

The Brassica oleracea Genome

Springer

Editors
Shengyi Liu
Oil Crops Research Institute
Chinese Academy of Agricultural Sciences, Ministry of Agriculture of PRC
Wuhan, China

Rod Snowdon
Department of Plant Breeding Research
Centre for BioSystems, Land Use
Justus Liebig University
Giessen, Germany

Chittaranjan Kole
Raja Ramanna Fellow, Government of India
ICAR-National Institute for Plant Biotechnology
Pusa, New Delhi, India

ISSN 2199-4781 ISSN 2199-479X (electronic)
Compendium of Plant Genomes
ISBN 978-3-030-31003-5 ISBN 978-3-030-31005-9 (eBook)
https://doi.org/10.1007/978-3-030-31005-9

This Springer imprint is published by the registered company Springer Nature Switzerland AG
The registered company address is: Gewerbestrasse 11, 6330 Cham, Switzerland

This book series is dedicated to my wife Phullara and our children Sourav and Devleena

Chittaranjan Kole

Preface to the Series

Genome sequencing has emerged as the leading discipline in the plant sciences coinciding with the start of the new century. For much of the twentieth century, plant geneticists were only successful in delineating putative chromosomal location, function, and changes in genes indirectly through the use of a number of 'markers' physically linked to them. These included visible or morphological, cytological, protein, and molecular or DNA markers. Among them, the first DNA marker, the RFLPs, introduced a revolutionary change in plant genetics and breeding in the mid-1980s, mainly because of their infinite number and thus potential to cover maximum chromosomal regions, phenotypic neutrality, absence of epistasis, and codominant nature. An array of other hybridization-based markers, PCR-based markers, and markers based on both facilitated construction of genetic linkage maps, mapping of genes controlling simply inherited traits, and even gene clusters (QTLs) controlling polygenic traits in a large number of model and crop plants. During this period, a number of new mapping populations beyond F2 were utilized and a number of computer programs were developed for map construction, mapping of genes, and for mapping of polygenic clusters or QTLs. Molecular markers were also used in the studies of evolution and phylogenetic relationship, genetic diversity, DNA fingerprinting, and map-based cloning. Markers tightly linked to the genes were used in crop improvement employing the so-called marker-assisted selection. These strategies of molecular genetic mapping and molecular breeding made a spectacular impact during the last one and a half decades of the twentieth century. But still they remained "indirect" approaches for elucidation and utilization of plant genomes since much of the chromosomes remained unknown and the complete chemical depiction of them was yet to be unraveled.

Physical mapping of genomes was the obvious consequence that facilitated the development of the 'genomic resources' including BAC and YAC libraries to develop physical maps in some plant genomes. Subsequently, integrated genetic-physical maps were also developed in many plants. This led to the concept of structural genomics. Later on, emphasis was laid on EST and transcriptome analysis to decipher the function of the active gene sequences leading to another concept defined as functional genomics. The advent of techniques of bacteriophage gene and DNA sequencing in the 1970s was extended to facilitate the sequencing of these genomic resources in the last decade of the twentieth century.

As expected, sequencing of chromosomal regions would have led to too much data to store, characterize, and utilize with the-then available computer software could handle. But the development of information technology made the life of biologists easier by leading to a swift and sweet marriage of biology and informatics, and a new subject was born—bioinformatics.

Thus, the evolution of the concepts, strategies, and tools of sequencing and bioinformatics reinforced the subject of genomics—structural and functional. Today, genome sequencing has traveled much beyond biology and involves biophysics, biochemistry, and bioinformatics!

Thanks to the efforts of both public and private agencies, genome sequencing strategies are evolving very fast, leading to cheaper, quicker, and automated techniques right from clone-by-clone and whole-genome shotgun approaches to a succession of second-generation sequencing methods. The development of software of different generations facilitated this genome sequencing. At the same time, newer concepts and strategies were emerging to handle the sequencing of the complex genomes, particularly the polyploids.

It became a reality to chemically—and so directly—define plant genomes, popularly called whole-genome sequencing or simply genome sequencing.

The history of plant genome sequencing will always cite the sequencing of the genome of the model plant Arabidopsis thaliana in 2000 that was followed by sequencing the genome of the crop and model plant rice in 2002. Since then, the number of sequenced genomes of higher plants has been increasing exponentially, mainly due to the development of cheaper and quicker genomic techniques and, most importantly, the development of collaborative platforms such as national and international consortia involving partners from public and/or private agencies.

As I write this preface for the first volume of the new series 'Compendium of Plant Genomes', a net search tells me that complete or nearly complete whole-genome sequencing of 45 crop plants, eight crop and model plants, eight model plants, 15 crop progenitors and relatives, and three basal plants is accomplished, the majority of which are in the public domain. This means that we nowadays know many of our model and crop plants chemically, i.e., directly, and we may depict them and utilize them precisely better than ever. Genome sequencing has covered all groups of crop plants. Hence, information on the precise depiction of plant genomes and the scope of their utilization are growing rapidly every day. However, the information is scattered in research articles and review papers in journals and dedicated Web pages of the consortia and databases. There is no compilation of plant genomes and the opportunity of using the information in sequence-assisted breeding or further genomic studies. This is the underlying rationale for starting this book series, with each volume dedicated to a particular plant.

Plant genome science has emerged as an important subject in academia, and the present compendium of plant genomes will be highly useful to both students and teaching faculties. Most importantly, research scientists involved in genomics research will have access to systematic deliberations on the plant genomes of their interest. Elucidation of plant genomes is of interest not only for the geneticists and breeders, but also for practitioners of an array of plant science disciplines, such as taxonomy, evolution, cytology, physiology, pathology, entomology, nematology, crop production, biochemistry, and obviously bioinformatics. It must be mentioned that information regarding each plant genome is ever-growing. The contents of the volumes of this compendium are, therefore, focusing on the basic aspects of the genomes and their utility. They include information on the academic and/or economic importance of the plants, description of their genomes from a molecular genetic and cytogenetic point of view, and the genomic resources developed. Detailed deliberations focus on the background history of the national and international genome initiatives, public and private partners involved, strategies and genomic resources and tools utilized, enumeration on the sequences and their assembly, repetitive sequences, gene annotation, and genome duplication. In addition, synteny with other sequences, comparison of gene families, and, most importantly, the potential of the genome sequence information for gene pool characterization through genotyping by sequencing (GBS) and genetic improvement of crop plants have been described. As expected, there is a lot of variation of these topics in the volumes based on the information available on the crop, model, or reference plants.

I must confess that as the series editor, it has been a daunting task for me to work on such a huge and broad knowledge base that spans so many diverse plant species. However, pioneering scientists with lifetime experience and expertise on the particular crops did excellent jobs editing the respective volumes. I myself have been a small science worker on plant genomes since the mid-1980s and that provided me the opportunity to personally know several stalwarts of plant genomics from all over the globe. Most, if not all, of the volume editors are my longtime friends and colleagues. It has been highly comfortable and enriching for me to work with them on this book series. To be honest, while working on this series I have been and will remain a student first, a science worker second, and a series editor last. And I must express my gratitude to the volume editors and the chapter authors for providing me the opportunity to work with them on this compendium.

I also wish to mention here my thanks and gratitude to the Springer staff, particularly Dr. Christina Eckey and Dr. Jutta Lindenborn for the earlier set of volumes and presently Ing. Zuzana Bernhart for all their timely help and support.

I always had to set aside additional hours to edit books beside my professional and personal commitments—hours I could and should have given to my wife, Phullara, and our kids, Sourav and Devleena. I must mention that they not only allowed me the freedom to take away those hours from them but also offered their support in the editing job itself. I am really not sure whether my dedication of this compendium to them will suffice to do justice to their sacrifices for the interest of science and the science community.

New Delhi, India Chittaranjan Kole

Preface

The diploid C genome brassica crop species *Brassica oleracea* is representative for one of the world's most important vegetable crops, with a broad species diversity spanning multiple cabbage varieties, cauliflower, broccoli, kohlrabi, Brussels sprouts, and kales. Furthermore, it is representative for the C-subgenome donor of the important allopolyploid oilseed crops *B. carinata* (Ethiopian mustard) and *B. napus* (rapeseed, canola), and it is also closely related to a vast range of inter-crossable, wild C genome *Brassica* species. Hence, the genome of *B. oleracea* represents a valuable and diverse resource for improvement of both vegetable and oilseed *Brassica* crops and for investigation of genetic diversity across diverse *Brassica* C genome species. Its close relationship to the model crucifer *Arabidopsis thaliana* and to the A-subgenome *Brassica* species *B. rapa* enables translational genomics and functional studies using genome resources from Arabidopsis and other brassicas. Its paleohexaploid genome structure also makes it a relevant and interesting model species for studies on the genomic impact of ancient polyploidization and subsequent gene loss during post-polyploidy diploidization processes.

The first two *B. oleracea* genome assembles were completed in 2014, capturing a high proportion of the gene space. In the interim, sequencing costs have dropped to the point where systematic pan-genome analysis is now possible using large-scale resequencing approaches, which have been applied to *B. oleracea* and its close relatives in studies investigating genetic and genomic diversity and their associations with traits. Moreover, long-read DNA sequencing methods and costs have also advanced to a point where they are now accessible even for complex crop genomes, where they offer considerable potential to vastly improve scaffold contiguity. Before publication of this book, the first long-read genome assembly of *B. oleracea* was completed in combination with optical mapping and short-read finishing.

The chapters included in this book provide an overview of the whole-genome sequencing, assembly, and annotation processes, along with physical map construction and description of the genome composition and organization. It also describes various aspects of the impact that the *B. oleracea* genome assembly has had and will continue to have, on genetics, genomics, and breeding. Genetic mapping, quantitative trait analysis, and gene cloning technologies are outlined, along with examples for analysis of important gene families and insight into how genome analysis can illuminate meso-polyploid genome evolution. Structural and functional genomic resources and the development of appropriate community databases are discussed in the framework of present and future genome *B. oleracea* resources and their implementation for genetic analyses and breeding.

It is becoming increasingly feasible to translate knowledge about genes and genomes into extremely targeted crop improvement, and *B. oleracea* was one of the first crop species in which CRISP-cas9 gene editing was successfully applied in a proof of concept. In future, comprehensive pan-genome analysis with unravel species-wide diversity and genome structural variation as a basis to translate population genomics into trait improvement by combining multi-level 'omics' approaches with biotechnological advances and classical breeding techniques. Genetic gain in breeding can benefit most greatly from acceleration of the time needed to advance generations or perform selection steps, and the *B. oleracea* genome lays a foundation for many different applications of genomics to improve efficiency and speed of key processes in this regard.

We are extremely grateful to the many chapter authors whose valuable expertise and contributions were essential for the completion of this book. We hope that this volume will serve as a useful reference for scholars, breeders, scientists, and students with an interest in *Brassica oleracea* and other brassica crops.

Wuhan, China — Shengyi Liu
Giessen, Germany — Rod Snowdon
Pusa, India — Chittaranjan Kole

Contents

Contributors

D. Babula-Skowrońska Institute of Plant Genetics, Polish Academy of Sciences, Poznań, Poland

Guusje Bonnema Plant Breeding, Wageningen University and Research, Wageningen, The Netherlands

Xiaohui Cheng Key Laboratory of Biology and Genetic Improvement of Oil Crops, The Ministry of Agriculture and Rural Affairs, Oil Crops Research Institute, Chinese Academy of Agricultural Sciences, Wuhan, China

J. Fidler Institute of Plant Genetics, Polish Academy of Sciences, Poznań, Poland

Weina Ge School of Life Sciences, North China University of Science and Technology, Tangshan, Hebei, China;
Center for Genomics and Computational Biology, North China University of Science and Technology, Tangshan, Hebei, China

Ning Guo Key Laboratory of Biology and Genetic Improvement of Horticultural Crops (North China), Beijing Vegetable Research Center, Beijing Academy of Agriculture and Forestry Sciences, National Engineering Research Center for Vegetables, The Ministry of Agriculture and Rural Affairs, Beijing, China

Arvind H. Hirani Department of Plant Science, University of Manitoba, Winnipeg, MB, Canada

Junyan Huang Key Laboratory of Biology and Genetic Improvement of Oil Crops, The Ministry of Agriculture and Rural Affairs, Oil Crops Research Institute, Chinese Academy of Agricultural Sciences, Wuhan, China

Hyun Hee Kim Bioscience Institute, Department of Chemistry and Life Science, Sahmyook University, Seoul, Republic of Korea

Chittaranjan Kole Raja Ramanna Fellow, Department of Atomic Energy, Government of India, ICAR-National Institute for Plant Biotechnology, New Delhi, India

Jonghoon Lee Department of Agriculture, Forestry and Bioresources, Plant Genomics & Breeding Institute, College of Agriculture and Life Sciences, Seoul National University, Seoul, Republic of Korea

Tianyu Lei School of Life Sciences, North China University of Science and Technology, Tangshan, Hebei, China;
Center for Genomics and Computational Biology, North China University of Science and Technology, Tangshan, Hebei, China

Genyi Li Department of Plant Science, University of Manitoba, Winnipeg, MB, Canada

Yuxian Li School of Life Sciences, North China University of Science and Technology, Tangshan, Hebei, China;
Center for Genomics and Computational Biology, North China University of Science and Technology, Tangshan, Hebei, China

Fan Liu Key Laboratory of Biology and Genetic Improvement of Horticultural Crops (North China), Beijing Vegetable Research Center, Beijing Academy of Agriculture and Forestry Sciences, National Engineering Research Center for Vegetables, The Ministry of Agriculture and Rural Affairs, Beijing, China

Lijiang Liu Key Laboratory of Biology and Genetic Improvement of Oil Crops, The Ministry of Agriculture and Rural Affairs, Oil Crops Research Institute, Chinese Academy of Agricultural Sciences, Wuhan, China

Shengyi Liu Key Laboratory of Biology and Genetic Improvement of Oil Crops, The Ministry of Agriculture and Rural Affairs, Oil Crops Research Institute, Chinese Academy of Agricultural Sciences, Wuhan, China

Jianxin Ma Department of Agronomy, Purdue University, West Lafayette, IN, USA

Yuxin Pan School of Life Sciences, North China University of Science and Technology, Tangshan, Hebei, China;
Center for Genomics and Computational Biology, North China University of Science and Technology, Tangshan, Hebei, China

Sampath Perumal Department of Agriculture, Forestry and Bioresources, Plant Genomics & Breeding Institute, College of Agriculture and Life Sciences, Seoul National University, Seoul, Republic of Korea;
Agriculture and Agri-Food Canada, Saskatoon, SK, Canada

Rod Snowdon Department of Plant Breeding, IFZ Research Centre for Biosystems, Land Use and Nutrition, Justus Liebig University, Giessen, Germany

Xiaoming Song School of Life Sciences, North China University of Science and Technology, Tangshan, Hebei, China;
Center for Genomics and Computational Biology, North China University of Science and Technology, Tangshan, Hebei, China

Chaobo Tong Key Laboratory of Biology and Genetic Improvement of Oil Crops, The Ministry of Agriculture and Rural Affairs, Oil Crops Research Institute, Chinese Academy of Agricultural Sciences, Wuhan, China

Nomar Espinosa Waminal Department of Agriculture, Forestry and Bioresources, Plant Genomics & Breeding Institute, College of Agriculture and Life Sciences, Seoul National University, Seoul, Republic of Korea;
Bioscience Institute, Department of Chemistry and Life Science, Sahmyook University, Seoul, Republic of Korea

Jinpeng Wang School of Life Sciences, North China University of Science and Technology, Tangshan, Hebei, China;
Center for Genomics and Computational Biology, North China University of Science and Technology, Tangshan, Hebei, China;
State Key Laboratory of Systematic and Evolutionary Botany, Institute of Botany, Chinese Academy of Sciences, Beijing, China;
University of Chinese Academy of Sciences, Beijing, China

Li Wang School of Life Sciences, North China University of Science and Technology, Tangshan, Hebei, China;
Center for Genomics and Computational Biology, North China University of Science and Technology, Tangshan, Hebei, China

Shenyun Wang Vegetable Research Institute, Jiangsu Academy of Agricultural Science, Nanjing, China

Xiyin Wang Plant Genome Mapping Laboratory, University of Georgia, Athens, GA, USA;
National Key Laboratory for North China Crop Improvement and Regulation, Agriculture University of Hebei, Baoding, China

Zhenyi Wang School of Life Sciences, North China University of Science and Technology, Tangshan, Hebei, China;
Center for Genomics and Computational Biology, North China University of Science and Technology, Tangshan, Hebei, China

Tae-Jin Yang Department of Agriculture, Forestry and Bioresources, Plant Genomics & Breeding Institute, College of Agriculture and Life Sciences, Seoul National University, Seoul, Republic of Korea

Jingyin Yu Key Laboratory of Biology and Genetic Improvement of Oil Crops, The Ministry of Agriculture and Rural Affairs, Oil Crops Research Institute, Chinese Academy of Agricultural Sciences, Wuhan, PR, China;
Boyce Thompson Institute for Plant Research, Cornell University, Ithaca, NY, USA

Min Yuan School of Life Sciences, North China University of Science and Technology, Tangshan, Hebei, China;
Center for Genomics and Computational Biology, North China University of Science and Technology, Tangshan, Hebei, China

Jiaqi Zhang School of Life Sciences, North China University of Science and Technology, Tangshan, Hebei, China;
Center for Genomics and Computational Biology, North China University of Science and Technology, Tangshan, Hebei, China

Lan Zhang School of Life Sciences, North China University of Science and Technology, Tangshan, Hebei, China;
Center for Genomics and Computational Biology, North China University of Science and Technology, Tangshan, Hebei, China

Meixia Zhao Department of Biology, Miami University, Oxford, OH, USA

N. Żyła Institute of Plant Genetics, Polish Academy of Sciences, Poznań, Poland

Abbreviations

3C	Chromosome conformation capture
AA	The diploid *Brassica rapa* A genome
AABB	The allotetraploid *Brassica juncea* AB genome
AACC	The allotetraploid *Brassica napus* AC genome
ABI	Advanced Biosciences Company
AFLP	Amplified fragment length polymorphism
A_n	Genome A derived from *Brassica napus*
ANOVA	Analysis of variance
A_r	Genome A derived from *Brassica rapa*
At	*Arabidopsis thaliana*
AUX	Gene AUX
BAC	Bacterial artificial chromosome
BB	The diploid *Brassica nigra* B genome
BBCC	The allotetraploid *Brassica carinata* BC genome
BBSRC	Biotechnology and Biological Sciences Research Council of Britain
BC	Backcross
BLAST	Basic local alignment sequence tool
Bo	*Brassica oleracea*
BoAP1a	*Brassica oleracea AP1a* gene
Boc1	*Brassica olearcea CACTA family 1*
BoCALa	*Brassica oleracea CALa* gene
BoCRB	Centromeric retrotransposon of *B. oleracea*
BoCS-lyase	*Brassica oleracea CS lyase* gene
BoCYP79F1	*Brassica oleracea CYP79F1* gene
BoDM1	*Brassica oleracea DM1* gene
BoGSL-ALK	*Brassica oleracea glucosinolate alkenyl synthase* gene
BoGSL-ELONG	*Brassica oleracea glucosinolate elongate* gene
BoGSL-PROa	*Brassica oleracea glucosinolate PROa* gene
BoGSL-PROb	*Brassica oleracea glucosinolate PROb* gene
BoGS-OH	*Brassica oleracea glucosinolate OH* gene
BolBase	*Brassica oleracea* genome database
BoS-GT	*Brassica oleracea GT* gene
BoSTR	Subtelomeric repeat of *B. oleracea* (variants a, b, c)
Bot1	*Brassica oleracea transposon 1*
bp	Base pair

BRAD	*Brassica rapa* Database
BrassicaTED	Brassica Transposable Elements Database
Brc1	*Brassica rapa CACTA family 1*
CACTA	A class II transposable element
Cap02-12	A *Brassica oleracea* cabbage cultivar
Cas9	CRISPR-associated protein 9
CC	The diploid *Brassica oleracea* C genome
CDPK	Calcium-dependent protein kinase
CDS	Coding DNA sequence
CENH3	Centromere-specific histone H3
Cent	Centromere
CentBo	Centromeric repeat of *B. oleracea* (variants 1 and 2)
CentBo1	A centromeric repeat sequence from *Brassica oleracea*
CentBr	A centromeric repeat sequence from *Brassica rapa*
CentBr1	A centromeric repeat sequence from *Brassica oleracea*
CentC	A centromeric repeat sequence from maize
CentO	A centromeric repeat sequence from rice
CFW	Cabbage Fusarium wilt
ChIP-seq	Chromatin immunoprecipitation sequencing
Chr	Chromosome
cM	CentiMorgan
C_n	Genome C derived from *Brassica napus*
CNV	Copy-number variant
CO	*Constans* gene
C_o	Genome C derived from *Brassica oleracea*
COG	Cluster of orthologous genes
COL	*Constans-Like* gene
CR	Centromere-specific retrotransposon
CR	Clubroot disease
CRB	Centromere-specific retrotransposon of *Brassica*
CRISPR	Clustered regularly interspaced short palindromic repeats
CRRs	A centromeric repeat sequence from rice
ctg	Contig
DDBJ	DNA Data Bank of Japan
DFS	Diversity foundation set
DH	Doubled haploid
DMS	Divergent mass selection
dsRNA	Double-stranded RNA
dw	Dry weight
EBI	European Bioinformatics Institute
EMBL	European Molecular Biology Laboratory
EST	Expressed sequence tag
ET	Ethylene
F_1	First filial generation
F_2	Second filial generation
F_3	Third filial generation

FAO	Food and Agriculture Organization (of the United Nations)
FISH	Fluorescent *in situ* hybridization
FKF1	*Kelch Repeat, F-Box 1* gene
FLC	*Flowering Locus C* gene
FPC	Fingerprint contig
FRI	*Frigida* gene
FT	*Flowering Locus T* gene
GBrowse	A genome browser software package
GBS	Genotyping by sequencing
GBS	Neoglucobrassicin
GER	Glucosinolates glucoerucin
GH3	Gene *GH3*
GI	*Gigantea* gene
GIV	Glucoiberverin
GMO	Genetically modified organism
GO	Gene ontology
GP	Genome proportion
GSL	Glucosinolate
GST	Glutathione S-transferase
GWA	Genomewide association
GWAS	Genomewide association study
hAT	A class of transposon
HDEM	A *Brassica oleracea* broccoli cultivar
HGBS	High content of orglucobrassicin
HGIB	Glucoiberin
Hi-C	A chromosome conformation capture method
HICF	High-information-content fingerprinting
HiSeq	A next-generation sequencing machine from the Illumina company
HPLC-DAD	High-performance liquid chromatography with a diode-array detector
HR	Hypersensitive response
HSIN	High content of sinigrin
HSR	Head splitting resistance
IAA	Gene *IAA*
InDel	Insertion-deletion
INRA	Institut national de la recherche agronomique (French National Institute for Agricultural Research)
IR	Illegitimate recombination
ITC	Isothiocyanates
JA	Jasmonic acid
JTT	Jones, Taylor, and Thorton model for maximal-likelihood tree construction
Ka	Non-synonymous base substitution rate
KEGG	Kyoto Encyclopedia of Genes and Genomes
Ks	Synonymous base substitution rate

KUpE	Potassium uptake efficiency
KUtE	Potassium utilization efficiency
LF	Least fractionated
LGBS	Glucobrassicin
LGIB	Low content of glucoiberin
LINE	Long interspersed nuclear elements
LRR	Leucine-rich repeat
LSIN	Sinigrin
LTR	Long terminal repeat
LTR-RT	LTR- retrotransposon
LWB	Large white butterfly
MAPK	Mitogen-activated protein kinase
MAS	Marker-assisted selection
MBGP	Multinational Brassica Genome Project
MEGA	Molecular Evolutionary Genetics Analysis
MF1	Moderately fractionated
MF2	Most fractionated
MGBS	Methoxyglucobrassicin
MinION	A long read sequencing system from Oxford Nanopore Technologies
MITE	Miniature inverted-repeat transposable element
MS	Mass spectrometry
MS	Microsoft company
MS-ESI	Mass spectrometry with electrospray ionization
mTE	Miniature transposable element
MY	Million years
MYA	Million years ago
NBS	Nucleolar binding site
NCBI	National Center for Biotechnology Information
NGBS	Glucosinolates neoglucobrassicin
NGS	Next-generation sequencing
nrDNA	Nuclear ribosomal DNA
nt	Nucleotide
NUpE	Nitrogren uptake efficiency
ONT	Oxford Nanopore Technology
PacBio	Pacific Biosciences company
PAML	Phylogenetic analysis by maximum likelihood
PAV	Presence-absence variant
PCR	Polymerase chain reaction
PCRBr	Pericentromere-specific retrotransposons of *Brassica*
PDS	*Phytoene Desaturase* gene
PIF	A class of transposon
PIN	Gene *PIN*
Pol III	RNA polymerase III
Pong	A class of transposon
PTGS	Posttranscriptional gene silencing
QTL	Quantitative trait locus

QTLs	Quantitative trait loci
QTL-seq	QTL bulk sequencing
RAPD	Random(ly) amplified polymorphic DNA
RC	Rolling circle
RCB	Rapid-cycling *Brassica*
RdDM	RNA-directed DNA methylation
rDNA	Ribosomal deoxyribose nucleic acid
RE	Repetitive element
RFLP	Restriction fragment length polymorphism
R-gene	Resistance gene
RNA-seq	RNA sequencing
ROS	Reactive oxygen species
RT	Retrotransposon
SA	Salicylic acid
SAR	Systemically activated resistance
SAUR	Gene *SAUR*
SCAR	Sequence-characterized amplified region
sgRNAs	Single guide RNA
SGS	Second-generation sequencing
SINE	Short interspersed elements
SINEs	Short interspersed repetitive elements
SiRNAs	Small interfering RNAs
SMRT	Single-molecule real-time sequencing
SNP	Single-nucleotide polymorphism
SRAP	Sequence-related amplified polymorphism
SSH	Suppression subtractive hybridization
SSR	*Sclerotinia* stem rot
SSR	Simple sequence repeat
STR	Subtelomeric tandem repeat
STS	Sequence-tagged site
TAA	Gene *TAA*
TAIR	The Arabidopsis Information Resource
Tc1	A class of transposon
TE	Transposable element
TGS	Third-generation sequencing
TGS	Transcriptional gene silencing
TIGR	The Integrated Genome Resource
TILLING	Targeting induced local lesions in genomes
TIR	Gene *TIR*
TIR-NBS-LRR	Toll/interleukin-1 receptor-nucleotide binding site-leucine rich repeat
TO1000	A *Brassica oleracea* kale cultivar
TPL	Gene *TPL*
TR	Tandem repeat
TRIM	Terminal-repeat retrotransposons in miniature
Ty1	A class of transposon
Ty3	A class of transposon

UDPG	Uridine diphosphate glucose
UR	Unequal homologous recombination
WGD	Whole-genome duplication
WGS	Whole-genome sequencing
WGT	Whole-genome triplication
WRKY	A transcription factor family
Xcc	*Xanthomonas campestris*
YUCCA	Gene *YUCCA*

Economic and Academic Importance of *Brassica oleracea* 1

N. Żyła, J. Fidler, and D. Babula-Skowrońska

Abstract

Brassica oleracea, which belongs to the family Brassicaceae, is an important vegetable crop worldwide. Domestication of this species has resulted in many diverse morphological types, such as cauliflower, broccoli, cabbage, Brussels sprouts, kohlrabi, and kale. Together with *B. rapa* and *B. nigra*, it is one of the components of the *Brassica* 'U' triangle (UN 1935). Furthermore, *B. oleracea*, as the source of the C genome, is one of the progenitor species of the allopolyploid crops *B. napus* (AC genomes) and *B. carinata* (BC genomes). It is also closely related to the model plant *Arabidopsis thaliana*. These features make *B. oleracea* essential for the study of polyploid evolution, and also for investigating new approaches to crop improvement.

1.1 Economic Importance

B. oleracea is grown as vegetable and ornamental plants as well as for forage. It is one of the oldest crops in the world. The wild cabbage, *B. oleracea* var. *oleracea* L., derives from southern and western Europe and dates back to the Middle Ages. This species is considered the ancestor of todays cultivars (Rakow 2004 for review), but a few authors have proposed *B. rupestris*, *B. montana,* and *B. cretica* to be the ancestors of kohlrabi, headed cabbages and kales, and broccoli and cauliflower, respectively (Yarnell 1956; Crisp and Tapsell 1993; Rakow 2004 for review). The cultivation of cabbage was probably initiated by the Celts—the Latin *Brassica* comes from the Celtic word 'brassic', meaning cabbage—but might have already been underway in ancient Greece and Rome. Other authors suggest that *B. oleracea* was first domesticated on the Atlantic coast and then primary cultivated plants were transferred to the Mediterranean region, where further morphological derivation occurred (Snogerup et al. 1990; Hodgkin 1995; Maggioni et al. 2018). It has also been suggested that any wild *Brassica* species with the C genome and a haploid complement of $n = 9$, which has the ability to intercross with cole crops, can be considered a progenitor (Zohary et al. 2012).

The breeding selection of wild *B. oleracea*, which aimed to strengthen various phenotypic traits, allowed a number of morphological forms to develop. Presently, the morphotypes of *B. oleracea* are divided into six groups. These are cabbages, characterized by folding of the leaves into 'heads' (var. *capitata* L.); kales (var. *acephala* L.); cauliflower and broccoli, with developed and enlarged inflorescences (var. *botrytis* and var. *italica* L., respectively); Chinese kale (var. *alboglabra* L.); kohlrabi, with an enlarged

N. Żyła · J. Fidler · D. Babula-Skowrońska (✉)
Institute of Plant Genetics, Polish Academy of Sciences, Strzeszyńska 34, 60-479 Poznań, Poland
e-mail: dbab@igr.poznan.pl

S. Liu et al. (eds.), *The Brassica oleracea Genome*, Compendium of Plant Genomes,
https://doi.org/10.1007/978-3-030-31005-9_1

basal stem (var. *gongylodes* L.); and Brussels sprouts, which have a proliferation of heading buds (var. *gemmifera* (DC.) Zenker) (Snogerup 1980). Additionally, a unique and diversified group of vegetable coles is tronchuda cabbage (*B. oleracea* var. *costata*), which is an important source of resistance to at least some diseases such as downy mildew, blackrot, white rust, blackleg, and turnip mosaic virus (Monteiro and Williams 1989; Dias et al. 1991; Ferreira et al. 1993; Santos et al. 1996; Sousa et al. 1997; Coelho et al. 1998). All species of *Brassica* cross easily, producing fertile offspring, which clearly indicates their close genetic relationship (Rakow 2004 for review). The morphological diversity of *B. oleracea* is a factor in its contribution to the human diet, and this, together with its potent anti-cancer and anti-oxidant properties, and an ability to adapt to various environmental conditions, ensures it represents a significant share of global crop production. According to the latest data, the value of cabbage crops produced worldwide in 2016 amounted to USD 19,103 million (https://faostat.fao.org/). Detailed data on area harvested, yield, and production, divided into continents, are collected in Table 1.1. These data show both the excellent adaptation of *Brassica* to different climatic conditions and the huge consumer demand for plants of the Brassicaceae family. A significant part of the production is carried out locally, but some large production centers export cabbage to many countries throughout the year (Rakow 2004 for review). The most important subspecies is the white cabbage *B. oleracea* var. *capitata*, which accounts for 25% of all Brassica cultivation. This is due to its relatively easy cultivation and associated high yields, together with low production costs and year-round popularity with consumers.

The *B. oleracea* varieties are mainly used as vegetable and fodder plants, although some subspecies can be grown as ornamental plants. Cabbage is also cultivated for its medicinal properties. Among others, these include phytotherapy treatment of inflammation and catarrh of the gastrointestinal tract, and treatment of kidney stones, as well as various respiratory diseases due to its bactericidal properties. The anti-cancer activity of *B. oleracea* has been confirmed in many studies and is thought to be due to its richness in aromatic and aliphatic glucosinolates (Cai et al. 2004; Matusheski et al. 2006; Singh et al. 2006). These features make it an important research object.

1.2 Academic Importance

B. oleracea offers many benefits for basic research. It is one of the three diploid *Brassica* species within the U triangle, which, since they are closely related, have the ability to hybridize with each other to form allopolyploid species. The C genome of *B. oleracea* comprises 630 Mb organized into nine chromosomes and is predicted to contain 45,758 protein-coding genes with an average transcript length of 1,761 bp (Liu et al. 2014). Its genome structure shows evidence of a whole-genome triplication event during the evolution of the Brassicaceae, as well as further chromosomal rearrangements resulting in genome reshuffling and gene losses (Babula et al. 2003; Lysak et al. 2005; Town et al. 2006; Ziolkowski et al. 2006; Liu et al. 2014). The availability of whole-genome sequences opens up new possibilities for *Brassica* research. For example, the genetic and evolutionary basis of the morphological diversity described above has been studied in detail (Thompson 1976; Snogerup 1980; Lázaro and Aguinagalde 1998; Song et al. 1990; Kalloo and Bergh 1993).

Due to their morphological diversity and specific agronomic traits, including rapid cycling and self-compatibility, the TO1000 line (kale-like morphotype, *B. oleracea* var. *alboglabra*) and the homozygous line 02–12 (heading cabbage, *B. oleracea* var. *capitata*) have been selected as models for genetic experiments. The TO1000 line has become a tool for the construction of a bacterial artificial chromosome (BAC)-based physical map (Wang et al. 2011), and for the derivation of a double-haploid mapping population (Iniguez-Luy et al. 2009) and a TILLING (targeting induced local lesions in

Table 1.1 Global production of cabbage and other *Brassica* by continent in 2017. These data include area harvested, yield, and production (https://faostat.fao.org/)

Area	Element	Value	Unit
Americas	Area harvested	76,681	ha
	Yield	315,729	hg/ha
	Production	242,103	tons
Asia	Area harvested	1,785,416	ha
	Yield	302,981	hg/ha
	Production	54,094,742	tons
Europe	Area harvested	373,060	ha
	Yield	306,252	hg/ha
	Production	11,425,045	tons
Australia and Oceania	Area harvested	3931	ha
	Yield	398,727	hg/ha
	Production	156,742	tons

genomes) population (Himelblau et al. 2009). Line 02–12 has been used as a parent in hybrid breeding (Wang et al. 2012). Subsequently, the homozygous double-haploid line TO1000DH3 (derived from *B. oleracea* cultivar TO1434) and line 02–12 were selected for reference genome sequencing (Liu et al. 2014; Parkin et al. 2014). Sequence data from two different genotypes can be used to understand the molecular basis of evolutionary processes and morphological variability in *B. oleracea*.

Presently, whole-genome sequences are available for the following species in the *Brassica* genus: *B. rapa, B. nigra, B. oleracea, B. napus* and *B. juncea* (Wang et al. 2011; Chalhoub et al. 2014; Liu et al. 2014; Parkin et al. 2014; Yang et al. 2016). Most of the data for these species are available at the Brassica Database (https://brassicadb.org/brad/). In the case of *B. oleracea*, in which two different genotypes have been sequenced, all comprehensive and partial genomic data are available in public databases. Thus, the whole-genome sequences together with comparative genomic information are deposited at the Brassica Database (https://brassicadb.org/brad/) and Bolbase (https://ocri-genomics.org/bolbase), while partial data including mapping populations, quantitative trait loci (QTLs), expressed sequence tags (ESTs) and cloned genes have been shared at the Brassica.info (https://www.brassica.info/resource/databases.php). Together, these data will allow a better understanding of the structural variability and evolution of the *B. oleracea* genome as well as the evolutionary relationships between closely related Brassicaceae species.

B. oleracea is an excellent model within the *Brassica* genus for the study of processes involved in polyploid evolution. *B. oleracea* (C genome) is closely related to diploid *B. rapa* (A genome), and together they form the amphidiploid *B. napus* (AC genomes). Comparative analyses based initially on genetic and physical mapping, and later on genome sequencing data, indicate that *B. oleracea* as well as other *Brassica* species are mesopolyploids, which underwent a hexaploidization event approximately 20–23 Mya (Babula et al. 2003; Ziolkowski et al. 2006; Beilstein et al. 2010; Parkin et al. 2014).

While the availability of two reference genomes of *B. oleracea* is useful for evolutionary studies, relatively few agronomically important genes have been identified so far using this resource. Furthermore, the sequenced genomes of two genotypes do not encompass the full gene repertoire found in the many variants of this species. Thus, there are many genes characterized by presence/absence variants (PAVs) and copy number variants, which are particularly relevant to phenotypic variability and agronomic traits, and which are specific for a given genotype. Lately, this problem has been overcome by the construction of the *B. oleracea* pangenome (Golicz et al. 2016), which has been developed on the basis of sequencing data for nine cultivated *B. oleracea* varieties and a wild relative, *B. macrocarpa*. Together, this pangenome includes 61,379 gene models, 18.7% of which represents variable genes (PAVs) related to disease resistance, flowering time, glucosinolate metabolism, and vitamin biosynthesis. The identification of these PAVs can be useful in association studies (Golicz et al. 2016).

Research on the function of genes and their participation in various developmental or signaling pathways has been markedly accelerated by the general applicability of the clustered regularly interspaced short palindromic repeats/ CRISPR-associated protein 9 (CRISPR/Cas9) mediated genome editing technology, which provides a precise tool for site-directed mutagenesis of genomes or individual genes of high complexity organisms. CRISPR consists of the RNA-guided endonuclease Cas9 and guide sequences designed to specifically position Cas9 at the target sequence. Several research groups have developed the CRISPR/Cas9 system for the *Brassica* genus (Lawrenson et al. 2015; Braatz et al. 2017; Kirchner et al. 2017; Yang et al. 2017; Murovec et al. 2018; Okuzaki et al. 2018; Sun et al. 2013; Lawrenson et al. 2019). Initial work showed its usefulness for a targeted mutation in *B. oleracea* and the stable transmission of such mutations through successive generations (Lawrenson et al. 2015). Subsequent work has elaborated a DNA-free genome editing protocol for site-directed mutagenesis using ribonucleoproteins (RNPs) in *B. oleracea* (Murovec et al. 2018). It has been tested inter alia in *B. oleracea* by introducing mutations in two target genes, the vernalization determinant FRIGIDA gene (*FRI*) and the phytoene desaturase gene (*PDS*), using two sgRNAs per gene. The frequency of mutations detected after transfection into cabbage protoplasts was in the range 0.09–2.25%. The sequence changes due to insertion or deletion were present 24 h after transfection and were gene- and locus-dependent. These studies indicate that CRISPR/Cas9 technology can be successfully used in *Brassica* species and we can assume that this method will be developed further in the coming years.

Research using whole-genome technologies, such as next generation sequencing (NGS), whole-genome profiling (RNA-seq), whole-genome analysis of protein-DNA interactions (ChIP-seq), and site-directed mutagenesis of genomes (CRISPR), show huge potential for *B. oleracea* breeding. New molecular technologies and easy availability of *Brassica* genomic information will facilitate and significantly accelerate the identification and then the functional characterization of individual genes responsible for important agronomic traits. Targets will include genes related to stress resistance and control of plant diseases, which may be identified by NGS-based methods such as genome-wide association studies and whole-genome single nucleotide polymorphism (SNP) genotyping. Such projects demonstrate the synergy of basic science and its applications in plant breeding.

References

Babula D, Kaczmarek M, Barakat A, Delseny M, Quiros CF, Sadowski J (2003) Chromosomal mapping of *Brassica oleracea* based on ESTs from *Arabidopsis thaliana*: complexity of the comparative map. Mol Genet Genom 268:656–665

Braatz J, Harloff H-J, Mascher M, Stein N, Himmelbach A, Jung C (2017) CRISPR-Cas9 targeted mutagenesis leads to simultaneous modification of different homoeologous gene copies in polyploid oilseed rape (*Brassica napus*). Plant Physiol 174:935–942

Cai Y, Luo Q, Sun M, Corke H (2004) Antioxidant activity and phenolic compounds of 112 traditional Chinese medicinal plants associated with anticancer. Life Sci 12:2157–2184

Chalhoub B, Denoeud F, Liu S, Parkin IA, Tang H, Wang X, Chiquet J, Belcram H, Tong C, Samans B, Corréa M, Da Silva C, Just J, Falentin C, Koh CS, Le Clainche I, Bernard M, Bento P, Noel B, Labadie K, Alberti A, Charles M, Arnaud D, Guo H, Daviaud C, Alamery S, Jabbari K, Zhao M, Edger PP et al (2014) Early allopolyploid evolution in the post-Neolithic *Brassica napus* oilseed genome. Science 345:950–953

Coelho P, Leckie D, Bahcevandziev K, Valério L, Astley D, Boukema I, Crute I, Monteiro AA (1998) The relationship between cotyledon and adult plant resistance to downy mildew (*Peronospora parasitica*) in *Brassica oleracea*. Acta Hort 459:335–342

Crisp P, Tapsell CR (1993) Cauliflower, *Brassica oleracea* L. In: Kalloo G, Bergh BO (eds) Genetic Improvement of Vegetable Crops. Pergamon Press, Oxford, pp 157–178

Dias JS, Ferreira ME, Williams PH (1991) Resistance of Portuguese coles to blackrot. Acta Hort 8:213–218

Ferreira ME, Dias JS, Mengistu A, Williams PH (1993) Screening of Portuguese cole landraces (*Brassica oleracea* L.) with *Leptosphaeria maculans* and *Xanthomonas campestris* pv. *campestris*. Euphytica 65:219–227

Golicz A, Bayer PE, Barker GC, Edger P, Kim H, Martinez PA, Kit C, Chan K, Severn-Ellis A, McCombie WR, Parkin IAP, Paterson AH, Pires JC, Sharpe AG, Tang H, Teakle GR, Town CD, Batley J, Edwards D (2016) The pangenome of an agronomically important crop plant *Brassica oleracea*. Nat Commun 7:13390

Himelblau E, Gilchrist EJ, Buono K, Bizzell C, Mentzer L, Vogelzang R, Osborn T, Amasino RM, Parkin IA, Haughn GW (2009) Forward and reverse genetics of rapid-cycling *Brassica oleracea*. Theor Appl Genet 118:953–961

Hodgkin T (1995) Cabbages, kales, etc. *Brassica oleracea* (Cruciferae). In: Smartt J, Simmonds JW (eds) Evolution of Crop Plants, 2nd edn. Longman Scientific & Technical, Harlow, pp 76–82.

Iniguez-Luy FL, Lukens L, Farnham MW, Amasino RM, Osborn TC (2009) Development of public immortal mapping populations, molecular markers and linkage maps for rapid cycling *Brassica rapa* and *B. oleracea*. Theor Appl Genet 120:31–43

Kalloo G, Bergh B (1993) (eds) Genetic Improvement of Vegetable Crops. Pergamon Press, Oxford and New York.

Kirchner TW, Niehaus M, Debener T, Schenk MK, Herde M (2017) Efficient generation of mutations mediated by CRISPR/Cas9 in the hairy root transformation system of *Brassica carinata*. PLoS ONE 12: e0185429

Lawrenson T, Shorinola O, Stacey N, Li C, Østergaard L, Patron N, Uauy C, Harwood W (2015) Induction of targeted, heritable mutations in barley and *Brassica oleracea* using RNA-guided Cas9 nuclease. Genome Biol 16:258

Lawrenson T, Hundleby P, Harwood W (2019) Creating targeted gene knockouts in *Brassica oleracea* using CRISPR/Cas9. Methods Mol Biol 1917:155–170

Lázaro A, Aguinagalde I (1998) Genetic diversity in *Brassica oleracea* L. and wild relatives ($2n$ = 18) using RAPD markers. Ann Bot 82:829–833

Liu S, Liu Y, Yang X, Tong C, Edwards D, Parkin IA, Zhao M, Ma J, Yu J, Huang S, Wang X, Wang J, Lu K, Fang Z, Bancroft I, Yang T, Hu Q, Wang X, Yue Z, Li H, Yang L, Wu J, Zhou Q, Wang W, King GJ, Pires JC, Lu C, Wu Z, Sampath P, Wang Z, Guo H, Pan S, Yang L et al (2014) The *Brassica oleracea* genome reveals the asymmetrical evolution of polyploid genomes. Nat Commun 5:3930

Lysak MA, Koch MA, Pecinka A, Schubert I (2005) Chromosome triplication found across the tribe *Brassiceae*. Genome Res 15:516–525

Maggioni L, von Bothmer R, Poulsen G, Lipman E (2018) Domestication, diversity and use of *Brassica oleracea* L., based on ancient Greek and Latin texts. Genet Resour Crop Evol 65: 137.

Matusheski NV, Swarup R, Juvik JA, Mithen R, Bennett M, Jeffery EH (2006) Epithiospecifier protein from broccoli (*Brassica oleracea* L. ssp. *italica*) inhibits formation of the anticancer agent sulforaphane. J Agri Food Chem 54:2069–2076

Monteiro AA, Williams PH (1989) The exploration of genetic resources of Portuguese cabbage and kale for resistance of several *Brassica* diseases. Euphytica 41:215–225

Murovec J, Gucek K, Bohanec B, Avbelj M, Jerala R (2018) DNA-free genome editing of *Brassica oleracea* and *B. rapa*protoplasts using CRISPR-Cas9 ribonucleoprotein Complexes. Front Plant Sci 9:1594.

OkuzakiA OT, Koizuka C, Kaneko K, Inaba M, Imamura J, Koizuk N (2018) CRISPR/Cas9-mediated genome editing of the fatty acid desaturase 2 gene in *Brassica napus*. Plant Physiol Biochem 131:63–69

Parkin IA, Koh C, Tang H, Robinson SJ, Kagale S, Clarke WE, Town CD, Nixon J, Krishnakumar V, Bidwell SL, Denoeud F, Belcram H, Links MG, Just J, Clarke C, Bender T, Huebert T, Mason AS, Pires JC, Barker G et al (2014) Transcriptome and methylome profiling reveals relics of genome dominance in the mesopolyploid *Brassica oleracea*. Genome Biol 15: R77

Rakow G (2004) Species Origin and Economic Importance of *Brassica*. In: Pua EC, Douglas CJ (eds) Biochnology in Agriculture and Forestry, Vol. 54 Brassica. Springer-Verlag, Berlin, Heidelberg

Santos MP, Dias JS, Monteiro AA (1996) Screening portuguese cole landraces for resistance to white rust (*Albugo candida* (Pers.) Kuntze). Act Hort 407: 453–459

Singh J, Upadhyay AK, Bahadur A, Singh B, Singh KP, Rai M (2006) Antioxidant phytochemicals in cabbage (*Brassica oleracea* L. var. *capitata*). Sci Hort 108:233–237

Snogerup S (1980) The wild forms of the *Brassica oleracea* group (2n = 18) and their possible relations to the cultivated ones. In: Tsunoda S, Hinata K, Gómez-Campo C (eds) *Brassica* Crops and Wild Allies. Japan Scientific Societies Press, Tokyo, Biology and Breeding, pp 121–132

Snogerup S, Gustafsson M, von Bothmer R (1990) Brassica sect. Brassica (Brassicaceae). I. Taxonomy and Variation. Willdenowia 19:271–365

Song K, Osborn TC, Williams PH (1990) *Brassica* taxonomy based on nuclear restriction fragment length polymorphisms (RFLPs) 3. Genome relationship in *Brassica* and related genera and the origin of *B. oleracea* and *B. rapa* (syn. *campestris*). Theor Appl Genet 79:497–506

Sousa ME, Dias JS, Monteiro AA (1997) Screening Portuguese cole landraces for resistance to seven indigenous downy mildew isolates. Sci Hort 68:49–58

Sun Z, Li N, Huang G, Xu J, Pan Y, Wang Z, Tang Q, Song M, Wang X (2013) Site-specific gene targeting using transcription activator-like effector (TALE)-based nuclease in *Brassica oleracea*. J Integr Plant Biol 55:1092–1103

Thompson KF (1976) Cabbages, kales, etc. *Brassica oleracea* (Cruciferae). In: Simmonds NW (ed) Evolution of Crop Plants. Longman, London, pp 49–52

Town CD, Cheung F, Maiti R, Crabtree J, Haas BJ, Wortman JR, Hine EE, Althoff R, Arbogast TS, Tallon LJ, Vigouroux M, Trick M, Bancroft I (2006) Comparative genomics of *Brassica oleracea* and *Arabidopsis thaliana* reveal gene loss, fragmentation, and dispersal after polyploidy. Plant Cell 18: 1348–1359

UN (1935) Genome analysis in *Brassica* with special reference to the experimental formation of *B. napus* and peculiar mode of fertilization. Jpn J Bot 7: 389–452

Wang GX, Tang Y, Yan H, Sheng XG, Hao WW, Zhang L, Lu K, Liu F (2011) Production and characterization of interspecific somatic hybrids between *Brassica oleracea* var. *botrytis* and *B. nigra* and their progenies for the selection of advanced pre-breeding materials. Plant Cell Rep 30:1811–1821

Wang W, Huang S, Liu Y, Fang Z, Yang L, Hua W, Yuan S, Liu S, Sun J, Zhuang M, Zhang Y, Zeng A (2012) Construction and analysis of a high-density genetic linkage map in cabbage (*Brassica oleracea* L.var. *capitata*). BMC Genomics 13:523.

Yarnell SH (1956) Cytogenetics of the vegetable crops. II. Crucifers. The Botanical Review 22:81–166

Yang J, Liu D, Wang X, Ji C, Cheng F, Liu B, Hu Z, Chen S, Pental D, Ju Y, Yao P, Li X, Xie K, Zhang J, Wang J, Liu F, Ma W, Shopan J, Zheng H, Mackenzie SA, Zhang M (2016) The genome sequence of allopolyploid *Brassica juncea* and analysis of differential homoeology gene expression influencing selection. Nat Genet 48:1225–1232

Yang H, Wu JJ, Tang T, Liu KD, Dai C (2017) CRISPR/Cas9-mediated genome editing efficiently creates specific mutations at multiple loci using one sgRNA in *Brassica napus*. Sci Rep 7:7489

Ziółkowski PA, KaczmarekM BabulaD, Sadowski J (2006) Genome evolution in *Arabidopsis/Brassica*: conservation and divergence of ancient rearranged segments and their breakpoints. Plant J 47:63–74

Zohary D, Hopf M, Weiss E (2012) Domestication of plants in the Old World, 4th edn. Oxford University Press, Oxford

Genetic Mapping, Quantitative Trait Analysis, and Gene Cloning in *Brassica oleracea*

2

Arvind H. Hirani and Genyi Li

Abstract

Among the cultivated Brassica species, *B. oleracea* is among the most genetically and morphologically diverse species. The modern crop forms of *B. oleracea* have been cultivated and selected for centuries together with the diverse range of wild forms found in this species and closely related species, which played important roles in creating variation in *B. oleracea* and other cultivated Brassica species. Utilization of molecular breeding tools has been a great advantage in *B. oleracea* vegetable breeding around the world. Construction of genetic maps based on molecular markers ranging from isozyme to single nucleotide polymorphism (SNP) markers has enabled mapping of various important genes and quantitative trait loci (QTL) for yield, quality, disease resistance, and earliness in *B. oleracea*. This chapter focused on milestones in the utilization of various genomics tools such as genetic map construction, gene and QTL (fine) mapping, molecular marker development and marker-assisted selection in *B. oleracea* vegetable crops for enhancing productivity, quality, and yield to meet ever increasing vegetable demand. Use of genetic resources such as male sterile systems, restoration of male sterility for hybrid seed production and exploration of heterosis in hybrid Brassica vegetable breeding is also discussed.

A. H. Hirani · G. Li (✉)
Department of Plant Science, University of Manitoba, Winnipeg, MB, Canada
e-mail: genyi.li@umanitoba.ca

2.1 Introduction

The Brassicaceae family is one of the most important plant families among the agricultural and horticultural crops. It comprises more than 3,700 species which are distributed among 338 genera (Al-Shehbaz et al. 2006; Franzke et al. 2011), and is considered as a cosmopolitan family which carries long history in the Mediterranean with diverse morph-types. Globally, Brassica crops are renowned for their edible oil and dietary rich various types of vegetable production including leafy, root, stem, bud, and floret. *Brassica* species are also being adopted for fodder and condiment production in certain parts of the world. Apart from delivering healthy vegetable oil, *Brassica* species are major suppliers of vegetables with importance sources of dietary fiber, vitamin C and other nutritional elements such as anti-cancer compounds to human health.

Brassica oleracea is one of the most important vegetable-producing species of the Brassicaceae family globally. Heading cabbage and cauliflower were the two Brassica vegetables

S. Liu et al. (eds.), *The Brassica oleracea Genome*, Compendium of Plant Genomes,
https://doi.org/10.1007/978-3-030-31005-9_2

most widely consumed in past decades, but recently other Brassica vegetables such as broccoli, kale, collards, and Brussels sprouts are also gaining in popularity as salad vegetables due to their high nutraceutical properties. Brassica vegetables, especially *B. oleracea* subspecies, have a high preference among consumers due to the presence of beneficial compounds such as glucosinolates, phenolic compounds, anthocyanins, and flavonoid antioxidants which can play important roles in fighting cancers and degenerative diseases.

Among all cultivated vegetable plant species, Brassica vegetables are considered to be the most consumed forms of vegetables, which reflects attraction of consumers due to their beneficial health properties such as plant secondary metabolites like glucosinolates and their derivatives, and a richness of antioxidants like carotenoids, tocopherols, phenolic acids, sterols, and other flavonoids, which play important roles in healthy diets. Globally, the Brassica vegetable growing area is under contentious expansion in more than two decades (Fig. 2.1). Global production of commercial Brassica vegetables was over 91 million tonnes from the cultivated area around 3.6 million hectares, with a gross production value of about US$ 27 billion in 2012 (FAO 2016).

In the Brassica genus, four species are cultivated for the different forms of vegetable production in various countries: *B. oleracea*, *B. rapa*, *B. juncea,* and *B. napus*. Among these four species, *B. oleracea* has been grown for different forms of vegetables such as leaves, stems, lateral buds, terminal flower buds and flowers and it has been widely adopted in different geographical regions globally. In *B. oleracea*, existence of great genetic diversity for a variety of morphological and quality traits produces distinct vegetables which allow consumers to have multiple choices. Various traits were selected as modified organs in *B. oleracea*, such as leafy forms (kale and curly kale), terminal leaf bud (cabbage), flower bud and stem (broccoli, collard greens or collards), lateral leaf bud (Brussels sprout), lateral meristem (kohlrabi or turnip cabbage or German turnip) and flower bud (cauliflower) (Fig. 2.2). *B. oleracea* is categorized into seven subspecies based on its plant parts being used as an edible vegetable (Table 2.1).

2.2 Integrity and Originality of the C-Genome in *Brassica* Species

The diploid progenitor species, *B. oleracea* (CC genome, $2n = 18$), *B. rapa* (AA genome, $2n = 20$), and *B. nigra* (BB genome, $2n = 16$) of the three amphidiploid Brassica species *B. napus* (AACC genome, $2n = 38$), *B. carinata* (AABB

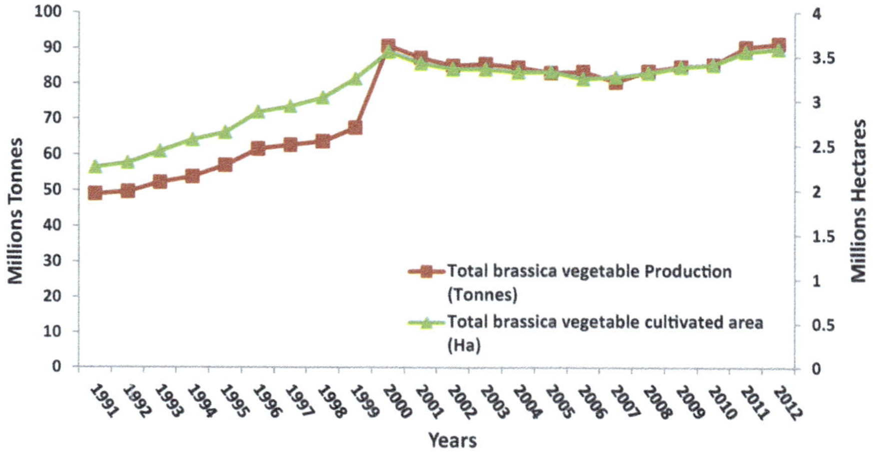

Fig. 2.1 Total production and area under cultivation of Brassica vegetables, globally, since 1991 (FAO 2016)

Fig. 2.2 Various forms for *B. oleracea* vegetables displayed existence of genetic and morphological diversity among *Brassica* species

Table 2.1 Subspecies of *B. oleracea* commonly cultivated for different forms of vegetable around the world

Subspecies	Name of the cultivated forms of species
B. oleraceaviridis	Kale
B. oleraceaacephala	Smooth-leafed kale, curly kale, collard greens, thousand-head kale, tree kale, marrow-stem kale
B. oleraceaalboglabra	Kai-lan or Chinese broccoli
B. oleracea botrytis	White-/purple- cauliflower, romanesco broccoli, broccoflower
B. oleraceacapitata	Green cabbage, red cabbage
B. oleraceagemmifera	Brussels sprouts
B. oleraceagongylodes	Kohlrabi
B. oleraceaitalica	Broccoli

genome, $2n = 36$), and *B. juncea* (BBCC genome, $2n = 34$) harbor a vast amount of genetic diversity. The progenitor species play a vital role in maintaining genetic diversity for characteristics related to basal morphology, physiology, and evolutionary biology as well as commercially important traits.

Brassica genomes possess distinguished characteristics of a high level of DNA methylation polymorphism (Salmon et al. 2008), chromosomal rearrangements (Quiros and Paterson 2004; Qiu et al. 2009), sustaining viability of embryos in wide hybridizations through interspecific, intergeneric crosses (Kaneko and Bang 2014), and existence of these kinds of characteristics may be due to the Brassica genomes which were derived from initial polyploidization followed by diploidization during separation process.

Comparative genome mapping of Arabidopsis with other Brassica species shows that one locus in the Arabidopsis genome is represented by three loci in the diploid Brassica genomes (Osborn et al. 1997; Lagercrantz 1998). The whole-genome triplication, followed by chromosome rearrangements occurred through fissions and fusions of different fragments that led to the shaping of Brassica ancestors. The Brassica genomes were evolved during speciation process, though each genome sustained the genomic structure of Arabidopsis with sustainable functional properties. The segmental genomes originated from ancestors which underwent genome triplication and subsequent structural reorganization, resulting in the establishment of the Brassica diploid genomes with single copy to multiple copies of genes. For example, four homologs of FLOWERING LOCUS C (*FLC*) could be identified in *B. rapa* (*BrFLC1-3*and *BrFLC5*), four in *B. oleracea* (*BoFLC1*, *BoFLC3-5;* Okazaki et al. 2007; Schranz et al. 2002), and five loci in *B. napus* denominated *BnFLC1-5* (Tadege et al. 2001).

2.3 Genetic Map Construction in *B. oleracea*

Genetic map construction is a key component for mapping qualitative and quantitative traits. Ultra-high-density linkage maps with even spanning of reliable molecular markers is important in molecular breeding applications including gene/quantitative trait locus (QTL) mapping, fine mapping, cloning and eventually integration of useful genes/QTL to novel breeding materials through marker-assisted selection. Currently, dozens of genetic maps have been developed in *B. oleracea* using different populations and molecular markers (Table 2.2). In the earlier biotechnology era, Arus and Orton (1983) initiated first isozyme markers for linkage map construction in *B. oleracea*. Subsequently accessibility of DNA molecular markers including hybridization-based restriction fragment length polymorphism (RFLP) and PCR-based simple sequence repeats (SSR), intersimple sequence repeats (ISSR), random amplified polymorphic DNA (RAPD), amplified fragment length polymorphism (AFLP), and sequence-related amplified polymorphism (SRAP) are explored by many research groups to create genetic maps from different accessions of *B. oleracea*. These genetic maps were extensively reviewed by (Quiros 2000; Quiros and Paterson 2004) with respect to types of molecular markers, type of segregation populations and their parents. The first fairly large genetic linkage map was constructed using 258 RFLP markers based on an F_2 population derived from cabbage and broccoli which covered 820 cM genetic distance of the C-genome (Slocum et al. 1990). Furthermore, Kianian and Quiros (1992) developed linkage map from four different F_2 populations, consisted 108 RFLP, isozyme and morphological markers covering about 747 cM in 11 linkage groups. Similarly, a genetic linkage map was constructed using 159 RFLP markers from an F_2 segregation population derived from a cross between cabbage and broccoli with the total coverage of 921 cM (Camargo 1996).

Subsequently a consensus map for the *B. oleracea* from two doubled haploid (DH) populations (broccoli X kale and cauliflower X Brussels sprouts) with about 550 RFLP, amplified fragment length polymorphism (AFLP), and simple sequence repeat (SSR) markers was constructed by Sebastian et al. (2000) and later on all the nine linkage groups were physically assigned to their respective chromosomes using fluorescence in situ hybridization (FISH) and new markers added to several linkage groups (Howell et al. 2002). In addition, these chromosomes were also used to align linkage groups of both the A and C genomes in *B. napus* (Lowe et al. 2004; Piquemal et al. 2005). Since the early 2000s, availability of Arabidopsis EST sequences opened the way for construction of several genetic linkage maps along with EST-based partial comparison of the Arabidopsis and *B. oleracea* genomes (Babula et al. 2003; Lan and Paterson 2000; Li et al. 2003a).

The first high-density genetic linkage map of *B. oleracea,* with 1,257 molecular markers including 247 cDNA-based SRAP markers spanning 703 cM, was constructed based on an F_2 population derived from a broccoli and cauliflower cross (Gao et al. 2007). In this genetic map, all the linkage groups were assigned to their respective chromosomes based on common markers used by Piquemal et al. (2005); (Sebastian et al. 2000), and also to the chromosomes of *A. thaliana*. Similarly, the second high-density genetic map was constructed with 1,227 markers and a DH population was derived from diverged cabbage lines, and this map covered a total genetic distance of 1,197.9 cM (Wang et al. 2012). Recently, a high-density genetic linkage map of *B. oleracea* was constructed with 4,103 GBS-SNP (genotype-by-sequencing SNP) in $F_{2:3}$

Table 2.2 Genetic linkage maps constructed using different molecular markers and segregating populations in *B. oleracea*

Parental crosses	Populations	Types of markers	Total markers	Map coverage	References
Cabbage inbred lines (C1176 X C1220)	F_2:3 progenies	GBS-SNP	4103	879.9	Lee et al. (2015a)
Broccoli var. VI-158 X Broccoli var. BNC	F_3 families	SNP array	547	948.1	Brown et al. (2014)
Cabbage lines: (C1184 X C1234)	F_2	SSR and SNP	265	1331.88	Izzah et al. (2014)
B. oleracea(*B. incana*) X *B. oleracea*	F_2	SRAP	267	1087.6	Mei et al. (2013)
Cabbage X Cabbage	DH	SSR and SNP	1227	1197.9	Wang et al. (2012)
Broccoli DH of cv. Green Duke X Marathon	F_2	AFLP and SSR	134	946.7	Walley et al. (2012)
Chinese cabbage X Broccoli	DH	RFLP and SSR	279	891.4	Iniguez-Luy et al. (2009)
Broccoli X Cauliflower	F_2	SRAP and SSR	1257	703	Gao et al. (2007)
Cauliflower X Cauliflower	F_2	AFLP and NBS	255	668.4	Gu et al. (2007)
Kale X Broccoli	F_2	RFLP	199	1226.3	Rocherieux et al. (2004)
Cabbage X Broccoli	F_2	AFLP, RAPD and SSR	405	731.9	Farinhó et al. (2004)
Chinese kale X Cabbage	F_2	RAPD	96	555.7	Chen and Wang (2002)
Chinese Kale X Broccoli Cauliflower X Brussels sprouts	2 DH	RFLP and AFLP	547	893	Sebastian et al. (2000)
Cabbage X Kale	F_2	RAPD, RFLP and isozyme	124	823.6	Moriguchi et al. (1999)
Collard X Cauliflower	F_2	RFLP	167	1738	Hu et al. (1998)
Cabbage X Chinese cabbage	F_2	RFLP, RAPD, STS, SCAR, phenotypic markers and isozymes	310	1606	Cheung et al. (1997)
Cabbage X Broccoli	DH	RFLP and AFLP	92	165	Voorrips et al. (1997)
Cabbage X Broccoli	F_2	RFLP and RAPD	159	921	Camargo et al. (1997)
Chinese kale X Broccoli	BC_1	RFLP, RAPD and isozyme	138	747	Ramsay et al. (1996)
B. oleracea var. (Green Duke X RMA12)	BC_1 BC_2	RFLP, RAPD and isozyme	126	1051	Kearsey et al. (1996)
Broccoli X Chinese kale	DH	PFLP	303	875	Bohuon et al. (1996)

(continued)

Table 2.2 (continued)

Parental crosses	Populations	Types of markers	Total markers	Map coverage	References
Cabbage X Broccoli	F_2	RFLP	112	1002	Camargo (1996)
Collard X Cauliflower. Collard X Broccoli, Kale X Cauliflower	3 F_2	RFLP and isozyme	108	747	Kianian and Quiros (1992)
Cabbage X Chinese cabbage	F_2	RFLP	201	1112	Landry et al. (1992)
Broccoli X Cabbage	F_2	RFLP	258	820	Slocum et al. (1990)
Broccoli X Brussels sprouts	F_2	Isozyme	–	–	Arus and Orton (1983)

progenies; this map covered a total genetic distance of 879.9 cM (Lee et al. 2015a). In addition, an F_2 derived F_3 family population was analyzed using the Illumina InfiniumSNP bead-array technology and 547 SNP markers were detected to construct a genetic map with a total coverage of 948.1 cM (Brown et al. 2014). Among all the linkage maps constructed in *B. oleracea*, molecular markers ranging from 92 to 4,103 were used. The GBS-SNP-based linkage map developed by Lee et al. (2015a) is the highest density genetic map with 4,103 GBS-SNP markers. The second highest density genetic map with 1,257 SRAP molecular markers was published almost a decade earlier (Gao et al. 2007). However, the chromosome lengths covered by the map ranged strongly, from 65 cM to 1,738 cM. Another highly dense map was constructed by Hu et al. (1998), with the base map adopted from the previous study carried out by Kianian and Quiros (1992). Generation of consensus map from the maps constructed in independent studies has been proven to be a difficult task in Brassica species in spite of using the same molecular marker systems or the same set of probes. There are several reasons, namely (i) the high duplication nature of the *B. oleracea* genome, multiple copies of genes or sequence fragments observed at the same locus or on different loci within the genome, (ii) different levels of polymorphism in the parental lines used for mapping population generation, (iii) probes or marker polymorphism varying from one population to another.

2.4 Gene/QTL Mapping in *B. oleracea*

Early adaptation, the subsequent long cultivation history and artificial selection of Brassica crops, especially among the vegetable Brassicas, has been commonly practiced globally. Recently, Brassica vegetable breeding has been supported by various biotechnology and genomics tools including genetic map construction, gene mapping, and fine mapping and cloning. These tools and technologies have been effectively employed for marker-assisted selection, genome-wide association (GWA) mapping, and genomic selection of elite germplasm and cultivars with superior quality and better agronomic performance. In Brassica vegetable breeding, seed yield is not an over-riding important factor, since higher seed yield is usually compensated by lower quality of vegetative organs. Brassica vegetable breeding tends to focus more on optimized quality characteristics such as tenderness, flavor, cosmetic appeal for good marketable quality, post-harvest shelf life and richness in antioxidants and nutritional elements.

2.5 Revealing Morphological and Quality Traits

In the *Brassica* genus, subspecies of *B. oleracea* exhibit a diverse range of leaf, stem, bud, and floral morphological characteristics, which are more interesting to plant geneticists to study and understand genes underpinning these characteristics and their role on plant productivity and adaptability. Vegetables belonging to *B. oleracea* possess unique morphological appealing attributes to the consumers, such as narrow stem of kale (*B. oleracea* spp. *medullosa*) and enlarged stem of kohlrabi (*B. oleracea* spp. *gongylodes*), enlarged inflorescence of broccoli (*B. oleracea* spp. *italic*) and cauliflower (*B. oleracea*spp. *botrytis*) and the enlarged apical bud of cabbage (*B. oleracea* spp. *capitata*) and lateral buds of Brussels sprouts (*B. oleracea* spp. *gemmifera*) (Lan and Paterson 2000).

Colorful vegetables and fruits have gained an increasing attention as nutraceutical foods due to their high level of pigmentation with potential health promoting and nutritional benefits. Among them, colorful Brassica vegetables are very eye-appealing vegetables available commercially. Recently, molecular identification of the '*Or*' gene conferred the edible orange color curds in cauliflower (*B. oleracea* sp. *botrytis*) and this gene is also involved in carotenoid biosynthesis, which is single semi-dominant spontaneous mutated gene in cauliflower (Crisp et al. 1975; Li et al. 2001b, 2003b; Lu et al. 2006). On the other hand, the recessive allele of the same gene '*or*' conferred orange inner leaf head in Chinese cabbage and also showed role in β-carotenoid biosynthesis (*B. rapa* ssp. *pekinensis*) (Feng et al. 2010; Zhang et al. 2013). The '*Pr*' gene encoding a R2R3 MYB transcription factor has the known role in regulation of purple color curds in purple cauliflower. Mutation in '*Pr*' gene conferred abnormal pattern of anthocyanin accumulation in purple cauliflower leaves and curds (Chiu et al. 2010).

Similarly, leaf color is an important trait for ornamental and vegetable kale (*B. oleracea* spp. *acephala)* breeding programs. Many kale cultivars are grown as flowering kale for their ornamental leaves such as violet interiors, blue, lavender, red, pink, and white (Wang et al. 2013). Studies reported that a single dominant gene '*Re*' determined the red color leaf in kale (Wang et al. 2013; Ren et al. 2015). Likewise, Zhang et al. (2011) reported that a MYB transcription factor induced under low temperature was involved in the biosynthesis of anthocyanin in purple kale. Recently, single semi-dominant gene, '*Pi*' was mapped for pink leaf color kale using SRAP and SSR molecular markers and six SRAP were converted into SCAR markers in an F_2 population derived from a cross between a white-leaved inbred line, W02–7, and a pink-leaved inbred line, P02–9 (Zhu et al. 2015).

Brassica vegetables are rich sources of secondary metabolites and flavonoids which enrich the nutritional properties of foods for healthy diets. Several secondary metabolites such as glucosinolates, carotenoids, tocopherols, ascorbate, phyloquinone, sterols, phenolic acids, and flavonoids are identified in different Brassica vegetable species. Glucosinolates are sulphur containing plant secondary metabolites, which are predominately found in the Brassicaceae family. The isothiocyanates and other hydrolytic breakdown products of glucosinolates are well known to have beneficial effects, such as reducing the incidence of cancer in humans and other mammals and the risk of degenerative diseases (Li et al. 2001a; Fahey and Talalay 1995; Fahey et al. 2001; Lin et al. 1998). Dietary groups of sulforaphane, isothiocyanates and indoles derived from Brassica vegetables are also known to stimulate apoptosis and confer protection against DNA damage in human colon cells (Bonnesen et al. 2001; Chiao et al. 2002). Therefore, genetic mechanisms involved in the biosynthesis of the glucosinolates in Brassica vegetables has been given more attention especially to manipulate beneficial glucosinolate compounds in Brassica vegetables for healthy diets. Previous studies reported that *B. oleracea* compared to *B. rapa* and *B. napus* displayed higher diversity for glucosinolate profile and quantity (Benderoth et al. 2008; Yang and Quiros

2010; Liu et al. 2014). Three major genes were identified and cloned in three crosses, broccoli X cauliflower, collard X broccoli, and collard X cauliflower in *B. oleracea* (Li et al. 2001a; Li and Quiros 2002, 2003). Among them, two genes, *BoGSL-ELONG* and *BoGSL-PRO*, which are homologous to *AtMAM-1* and *AtMAM-2* of Arabidopsis, respectively, are responsible for the side chain elongation of aliphatic glucosinolates (Li and Quiros 2002; Gao et al. 2006). The third gene, *BoGSL-ALK,* which is homologous to *AtAOP2* gene of Arabidopsis, is responsible for side chain modification and production of alkenyl glucosinolate (Li and Quiros 2003). In a subsequent study, together with above mentioned three major genes, 11 genes, *BoGSL-ALK, BoGSL-ELONG, BoGSL-PROa, BoGSL-PROb, BoCS-lyase, BoGS-OH, BoCYP79F1, BoS-GT, BoDM1, BoCALa*and*BoAP1a* were mapped in an F_2 segregating population derived from a cross between broccoli 'Early-Big' and cauliflower 'An-Nan Early' in *B. oleracea* (Gao et al. 2007). Furthermore, 82 QTLs were identified which explained a broad range of variability of individual and total glucosinolate content, among these QTLs, three genes controlled aliphatic glucosinolates and four genes controlled indole glucosinolate biosynthesis in a DH population of *B. oleracea* (Sotelo et al. 2014b).

Brassica vegetables are known for their richness in antioxidants such as phenolic compounds. Flavonoids and hydroxycinnamic acids are the widespread groups of polyphenols found in Brassica vegetables that scavenge reactive oxygen species and help release physiological stresses (Podsędek 2007). Sotelo et al. (2014a) reported 7 and 6 QTLs for antioxidant activity related to phenolics, anthocyanins and carotenoids in leaves and flower buds, respectively, in a DH population of *B. oleracea*. Likewise, 33 QTLs were found for total phenolic content and its individual components in leaves, flower buds and seeds in a DH population of *B. oleracea*, genomic regions detected were organ specific for total phenolic compounds or individual compound (Francisco et al. 2015).

2.6 Disease-Resistance Gene/QTL Mapping

Diseases caused by fungi, bacteria, oomycetes, viruses, and protists have been one of the most important factors reducing crop yield and causing quality losses ever since mankind initiated cultivation of plant species. Adaptation and continuous cultivation of major crops including cereal, oilseed, and vegetable crops have been immensely affected by diseases in different geographical regions of the globe. Brassica vegetable crops are the group of crops which have been greatly affected by various fungal, bacterial, and viral diseases. These diseases not only deteriorate quality of vegetable produce but also make sustainable production of these crops more challenging and unprofitable.

Fungal diseases such as blackleg (caused by *Leptosphaeria maculans/Phoma lingam*) Sclerotinia stem rot (*Sclerotiniasclerotiorum*), white mold or watery soft rot are the most important diseases in Brassica vegetables. These fungal diseases significantly deteriorate quality of the crops. Integrated management practise is an ideal management approach together with resistance breeding; however, quantitative inheritance nature of resistance is the biggest challenge for resistance breeding of these diseases. Blackleg disease can infect most of the Brassica species including all forms of vegetables belonging to *B. oleracea*. In 1925, infectious ascospores of *L. maculans* were firstly found on *B. oleracea* in Western Australia, mainly due to a lack of race-specific resistance genes in the C-genome of *Brassica* species. Later in 1970s, significant and widespread damage was caused by blackleg in Brassica cropping areas of Australia (Srivastava et al. 2010). The majority of race-specific resistance genes against blackleg disease have been mapped in the A-genome or B-genome of Brassica species, however, none of the genes have been mapped on the C-genome (Raman et al. 2013). Srivastava et al. (2010) reported molecular modeling and in silico docking methods to develop potential resistance-inducing candidate peptides against *L. maculans* in *B. oleracea*, however their applicability in Brassica vegetable breeding is not feasible.

Sclerotinia stem rot caused by *Sclerotinia sclerotiorum* is a non-host-specific and necrotizing fungus which infects over 400 plant species, including several important crops such as rapeseed, soybean and sunflower. Genetic sources with a high level of resistance to this disease are remained to be searched in *Brassica* species, it may be because of its complex quantitative resistance and the resistance is highly influenced by environment factors such as temperature and humidity. Recently, wild forms of *B. oleracea* such as *B. ruestris*, *B. incana*, *B. insularis* and *B. villosa* showed a high level of resistance to Sclerotinia stem rot by leaf and stem inoculation assay (Mei et al. 2010). These wild-type sources of *B. oleracea* could be a good source of resistance to Sclerotinia stem rot in other Brassica species. In a further study, a high level of resistance source, *B. incana* was used for QTL mapping with the susceptible parent *B. oleracea* var. *alboglabra*, an F_2 segregating population was used and 12 QTLs for leaf and 6 QTLs for stem resistance were found in two controlled condition experiments. Among them, 2 major QTLs for leaf and stem were on chromosome C09 (Mei et al. 2012).

Clubroot caused by protist *Plasmodiophora brassicae* is another devastating root disease in *Brassica* species found in almost all the *Brassica* species growing countries in the world. Clubroot disease significantly reduces crop yield and quality not only in Brassica vegetables but also in canola/rapeseed. In resistance gene and QTL mapping studies, at least eight race-specific resistance genes have been identified in *B. rapa* and molecular markers were developed for marker-assisted selection (Hirai et al. 2004; Piao et al. 2004; Suwabe et al. 2006; Sakamoto et al. 2008; Kato et al. 2012; Gao et al. 2014). In contrast to *B. rapa*, race-specific complete resistance has not been identified yet in *B. oleracea* (Chiang and Crete 1970; Piao et al. 2009; Hirani and Li 2015). However, several QTL mapping studies were carried out and reported QTLs in different mapping populations in *B. oleracea*. Recently, a genetic map constructed using 4,103 SNP markers was generated using GBS in $F_{2:3}$ progenies from clubroot-resistant and clubroot-susceptible *B. oleracea* lines. Two and one QTL(s) respectively were identified for *P. brassicae* races 2 and 9 (Lee et al. 2015a). Similarly, two major QTLs and three minor QTLs were found in F_3 progenies derived from a cross between a clubroot-resistant DH line (Anju) and a clubroot-susceptible DH line (GC) in *B. oleracea* (Nagaoka et al. 2010). In addition, there are several studies on clubroot resistance in *B. oleracea*(Dias et al. 1993; Voorrips and Visser 1993; Voorrips 1995; Voorrips and Kanne 1997), however, in most instances recessive inheritance of clubroot resistance was observed. This resistance cannot be effectively introgressed into other *Brassica* species for clubroot resistance breeding.

Vegetable crops belonging to *Brassica* species are also affected by various bacterial diseases. Among them, black rot caused by the bacterium *Xanthomonas campestris* pv. *campestris* has been reported to be a withering disease in many areas of the world (Singh et al. 2011b). This disease has a wide geographical distribution and is destructive to most *B. oleracea* vegetables, causing 50–65% yield and quality losses due to lack of resistance sources (Williams 1980). The pathogen invades through tissues of vascular bundle and progresses in V-shape chlorotic lesions, as a result leaf chlorosis at margins, necrosis and darkening of leaf veins and vascular tissues in the stem are the main disease symptoms. Resistance breeding to this disease is mainly hindered by two major factors in *B. oleracea*, namely (i) the quantitative nature of the resistance and (ii) the emergence of new races of the pathogen that overcome host resistance. Among the races identified so far, races 1 and 4 are aggressive and cause black rot disease in *B. oleracea* crops. A few studies have been performed for black rot resistance QTL mapping. For instance, Camargo et al. (1995) identified two QTLs on linkage groups C01 and C09 for both young and adult plant resistance in F_3 families derived from a cross between resistant cabbage and susceptible broccoli. Likewise, two major QTLs on linkage groups C02 and C09 and two minor QTLs on linkage groups C03 and C07 were identified in F_3 families derived from a cross between susceptible broccoli line, Green Commet P09 and resistant cabbage line, Reiho P01 (Doullah et al. 2011). Additionally, three QTLs were detected on linkage groups C02, C04, and C05 using EST-based SNP markers in the F_2 mapping population derived from a cross between resistant cabbage 'Early Fuji' and susceptible broccoli line (Kifuji et al. 2012). Furthermore, three QTLs were found on linkage groups C05, C08, and C09 for black rot resistance in a cross of susceptible DH line (GC P09) with a resistant DH line (Reiho P01) (Tonu et al. 2013). Recently, through whole genome resequencing of two black rot resistant and susceptible cabbage lines and genome-wide SNP identification using the reference *B. oleracea* genome, one major QTL and three minor QTLs in $F_{2:3}$ progenies were identified (Lee et al. 2015b).

2.7 Genetics of Heterosis and Hybrid Vigor in *B. oleracea*

Existence of genetic diversity has many advantages to improve productivity, quality and performance of crops under different biotic and abiotic stresses. Higher genetic diversity, especially in *B. oleracea*, provides opportunities to enhance combining ability of various beneficial traits together with increasing heterosis in hybrid cultivar development. Dey et al. (2013) reported evaluation of 40 hybrids derived from five cytoplasmic male sterile (CMS) lines and eight male fertile lines for heterosis and combining ability for vitamins and plant pigmentation in cauliflower. The study revealed that a CMS line had good general combining ability with all tester lines for ascorbic acid, anthocyanin content in cauliflower which provide scopes of increasing these pigments in hybrid cauliflower. Similarly, combining ability of 55 hybrids derived from five female lines and 11 male tester lines were carried out for mineral elements such as iron, zinc, copper, manganese, potassium, and calcium. The results indicated that specific combining ability was more important than general combining ability for mineral elements in cabbage head, and multiple crossing design was needed to increase all the mineral elements in hybrid (Singh et al. 2011a). Likewise, heterosis for earliness of head formation was evaluated through diallel analysis of four developmental traits: leaf position at which the head formation started, length of the 5th wrapper leaf, width of the 15th wrapper leaf and leaf shape index of width and length in cabbage, and the study found high broad and narrow sense heritabilities together with large additive and average dominant effects (Tanaka and Niikura 2006).

Earlier, hybridization and production of hybrids generally relied on self-incompatibility systems that were genetically complex and difficult to work with. Commercial hybrid seed production has been simplified by using more stable CMS systems (Renard et al. 1992). During the last two decades, noticeable progress has been made in the adaptation of Brassica vegetable hybrids (F_1) using different types of male sterility systems, with an objective of improving productivity, quality, earliness, and uniformity. The first male sterility in *B. oleracea* was transferred from *R. sativus* through repeated backcrosses with radish cytoplasm, in which *R. sativus* was used as female and *B. oleracea* was male parent (Ogura 1968; Bannerot et al. 1974). Subsequently, three Ogura-CMS lines were selected after BC_7 generations with superior commercial quality traits in cauliflower with one line exhibited highest general combining ability for curd yield and curd length and second line exhibited high specific combining ability for earliness. Varying degrees of heterosis were observed for yield related traits in cauliflower (Dey et al. 2011). Another CMS was introgressed from *B. rapa* to *B. oleracea* through protoplast fusion and subsequent shoot regeneration, in which mtDNA was strongly associated with male sterility and complete male sterility was observed without adverse effect in *B. oleracea* genetic background (Cardi and Earle 1997). Similarly, alloplasmic CMS and restorer lines were developed in *B. oleracea* through repeated backcrossing of *B. oleracea* as male parent with

Diplotaxis muralis as female parent, creating the so called *mur* CMS system. In the *mur* CMS system, chimeric mitochondrial gene *orf72* was located downstream of normal *rps7*and contained a portion of *atp9* (Shinada et al. 2006).

The combination of reliable and widely adaptable pollination systems with high heterosis for commercial traits is an important prerequisite for commercial hybrid Brassica vegetable breeding. New genome tools including SNP arrays, transcriptome profiles, and bioinformatics resources can be employed to evaluate heterotic combinations of different gene pools to enhance general combining and specific combining ability of hybrid breeding in Brassica vegetables.

2.8 Outlook

In summary, molecular tools and techniques such as molecular markers, QTL mapping, fine mapping and cloning of genes for useful traits, and marker-assisted selection have fostered the classical breeding process for novel cultivar development in *B. oleracea* crops. Significant progress has been made in overall Brassica vegetable quality improvement, although current and future challenges such as biotic and abiotic stresses, along with the need for increased yield to provide nutritious vegetable foods for the increasing world population, need to be addressed as high-priority objectives of vegetable breeding. The whole-genome sequencing data and annotation information available for the *B. oleracea* genome enable effectively implementation of genomics in Brassica vegetable crop improvement to enhance quality and productivity.

References

Al-Shehbaz IA, Beilstein MA, Kellogg EA (2006) Systematics and phylogeny of the *Brassicaceae* (Cruciferae): an overview. Plant Syst Evol 259(2–4):89–120. https://doi.org/10.1007/s00606-006-0415-z

Arus P, Orton T (1983) Inheritance and linkage relationships of isozyme loci in *Brassica oleracea*. J Hered 74:405–412

Babula D, Kaczmarek M, Barakat A, Delseny M, Quiros CF, Sadowski J (2003) Chromosomal mapping of *Brassica oleracea* based on ESTs from *Arabidopsis thaliana*: complexity of the comparative map. Mol Genet Genom 268(5):656–665. https://doi.org/10.1007/s00438-002-0782-2

Bannerot H, Boulidard L, Cauderon Y, Tempe J (1974) Transfer of cytoplasmic male sterility from *Raphanus sativus* to *Brassica oleracea*. Eucarpia meeting on Cruciferae. Dundee, Scotland, pp 52–54

Benderoth M, Pfalz M, Kroymann J (2008) *Methylthioalkylmalate synthases*: genetics, ecology and evolution. Phytochem Rev 8(1):255–268. https://doi.org/10.1007/s11101-008-9097-1

Bohuon EJR, Keith DJ, Parkin IAP, Sharpe AG, Lydiate DJ (1996) Alignment of the conserved C genomes of *Brassica oleracea* and *Brassica napus*. Theor Appl Genet 93(5–6):833–839. https://doi.org/10.1007/BF00224083

Bonnesen C, Eggleston IM, Hayes JD (2001) Dietary indoles and isothiocyanates that are generated from cruciferous vegetables can both stimulate apoptosis and confer protection against DNA damage in human colon cell lines. Cancer Res 61(16):6120–6130

Brown A, Yousef G, Chebrolu K, Byrd R, Everhart K, Thomas A, Reid R, Parkin IP, Sharpe A, Oliver R, Guzman I, Jackson E (2014) High-density single nucleotide polymorphism (SNP) array mapping in *Brassica oleracea*: identification of QTL associated with carotenoid variation in broccoli florets. Theor Appl Genet 127(9):2051–2064. https://doi.org/10.1007/s00122-014-2360-5

Camargo LEA (1996) Mapping RFLP and quantitative trait loci in *Brassica oleracea*. Theor Appl Genet 88:973–980

Camargo LEA, Williams PH, Osborn TC (1995) Mapping of quantitative trait loci controlling resistance of *Brassica oleracea* to *Xanthomonas campestris* pv. *campestris* in the field and greenhouse. Phytopathology 85:1296–1300

Camargo LE, Savides L, Jung G, Nienhuis J, Osborn TC (1997) Location of the self-incompatibility locus in an RFLP and RAPD map of *Brassica oleracea*. J Hered 88(1):57–60

Cardi T, Earle DE (1997) Production of new CMS *Brassica oleracea* by transfer of 'Anand' cytoplasm from *B. rapa* through protoplast fusion. Theor Appl Genet 94 (2):204–212. https://doi.org/10.1007/s001220050401

Chen S, Wang X (2002) Construction a RAPDs linkage map between Chinese kale and cabbage. Acta Hort Sin 29(3):229–232

Cheung WY, Champagne G, Hubert N, Landry BS (1997) Comparison of the genetic maps of *Brassica napus* and *Brassica oleracea*. Theor Appl Genet 94(5):569–582. https://doi.org/10.1007/s001220050453

Chiang MS, Crete R (1970) Inheritance of clubroot resistance in cabbage (*Brassica oleracea* L Var *Capitata*). Can J Genet Cytol 12(2):253–256

Chiao JW, Chung FL, Kancherla R, Ahmed T, Mittelman A, Conaway CC (2002) Sulforaphane and its metabolite mediate growth arrest and apoptosis in human prostate cancer cells. Intl J Oncol 20(3):631–636

Chiu LW, Zhou X, Burke S, Wu X, Prior RL, Li L (2010) The purple cauliflower arises from activation of a MYB transcription factor. Plant Physiol 154(3):1470–1480. https://doi.org/10.1104/pp.110.164160

Crisp P, Walkey DGA, Bellman E, Roberts E (1975) A mutation affecting curd colour in cauliflower (*Brassica oleracea* L. var. *Botrytis* DC). Euphytica 24(1):173–176. https://doi.org/10.1007/bf00147182

Dey SS, Sharma SR, Bhatia R, Parkash C, Barwal RN (2011) Superior CMS (Ogura) lines with better combining ability improve yield and maturity in cauliflower (*Brassica oleracea* var. *botrytis*). Euphytica 182(2):187–197. https://doi.org/10.1007/s10681-011-0425-y

Dey SS, Singh N, Bhatia R, Parkash C, Chandel C (2013) Genetic combining ability and heterosis for important vitamins and antioxidant pigments in cauliflower (*Brassica oleracea* var. *botrytis* L.). Euphytica 195 (2):169–181. https://doi.org/10.1007/s10681-013-0981-4

Dias JS, Ferreira ME, Williams PH (1993) Screening of Portuguese cole landraces (*Brassica oleracea* L.) with *Peronospora parasitica* and *Plasmodiophora brassicae*. Euphytica 67(1–2):135–141. https://doi.org/10.1007/bf00022736

Doullah M, Mohsin G, Ishikawa K, Hori H, Okazaki K (2011) Construction of a linkage map and QTL analysis for black rot resistance in *Brassica oleracea* L. Intl J Natural Sci 1(1):6. https://doi.org/10.3329/ijns.v1i1.8591

Fahey JW, Talalay P (1995) The role of crucifers in cancer chemoprotection. In: Gustine DL, Flores HE (eds) Phytochemicals and Health. American Society of Plant Physiologists, Rockville, MD, pp 87–93

Fahey JW, Zalcmann AT, Talalay P (2001) The chemical diversity and distribution of glucosinolates and isothiocyanates among plants. Phytochemistry 56(1):5–51

FAO (2016) Food and Agriculture Organization of the United Nations-statistical database. http://faostat.fao.org/site/567/DesktopDefault.aspx?PageID=567#ancor . Accessed Jan 27 2016

Farinhó M, Coelho P, Carlier J, Svetleva D, Monteiro A, Leitão J (2004) Mapping of a locus for adult plant resistance to downy mildew in broccoli (*Brassica oleracea* convar. *italica*). Theor Appl Genet 109 (7):1392–1398. https://doi.org/10.1007/s00122-004-1747-0

Feng H, Li Y, Liu Z, Liu J (2010) Mapping of *or*, a gene conferring orange color on the inner leaf of the Chinese cabbage (*Brassica rapa* L. ssp. *pekinensis*). Mol Breed 29(1):235–244. https://doi.org/10.1007/s11032-010-9542-x

Francisco M, Ali M, Ferreres F, Moreno-Fernández DA, Velasco P, Soengas P (2015) Organ-specific quantitative genetics and candidate genes of phenylpropanoid metabolism in *Brassica oleracea*. Front Plant Sci 6. https://doi.org/10.3389/fpls.2015.01240

Franzke A, Lysak MA, Al-Shehbaz IA, Koch MA, Mummenhoff K (2011) Cabbage family affairs: the evolutionary history of *Brassicaceae*. Trends Plant Sci 16(2):108–116. https://doi.org/10.1016/j.tplants.2010.11.005

Gao F, Hirani AH, Liu J, Liu Z, Fu G, Wu C, McVetty PBE, Li G (2014) Fine mapping a clubroot resistance locus in chinese cabbage. J Amer Soc Hort Sci 139(3):247–252

Gao M, Li G, Potter D, McCombie WR, Quiros CF (2006) Comparative analysis of *methylthioalkylmalate synthase (MAM)* gene family and flanking DNA sequences in *Brassica oleracea* and *Arabidopsis thaliana*. Plant Cell Rep 25:592–598. https://doi.org/10.1007/s00299-005-0078-1

Gao M, Li G, Yang B, Qiu D, Farnham M, Quiros C (2007) High-density *Brassica oleracea* linkage map: identification of useful new linkages. Theor Appl Genet 115(2):277–287. https://doi.org/10.1007/s00122-007-0568-3

Gu Y, Zhao QC, Sun DL, Song WQ (2007) Construction of genetic linkage map and localization of NBS-LRR like resistance gene analogues in cauliflower (*Brassica oleracea* var. *botrytis*). Hereditas 29(6):751–757

Hirai M, Harada T, Kubo N, Tsukada M, Suwabe K, Matsumoto S (2004) A novel locus for clubroot resistance in *Brassica rapa* and its linkage markers. Theor Appl Genet 108(4):639–643. https://doi.org/10.1007/s00122-003-1475-x

Hirani AH, Li G (2015) Understanding the genetics of clubroot resistance for effectively controlling this disease in *Brassica* Species. In: El-Shemy H (ed) Plants for the future. InTech. https://doi.org/10.5772/60936

Howell EC, Barker GC, Jones GH, Kearsey MJ, King GJ, Kop EP, Ryder CD, Teakle GR, Vicente JG, Armstrong SJ (2002) Integration of the cytogenetic and genetic linkage maps of *Brassica oleracea*. Genetics 161(3):1225–1234

Hu J, Sadowski J, Osborn TC, Landry BS, Quiros CF (1998) Linkage group alignment from four independent (*Brassica oleracea*) RFLP maps. Genome 41 (2):226–235

Iniguez-Luy F, Lukens L, Farnham M, Amasino R, Osborn T (2009) Development of public immortal mapping populations, molecular markers and linkage maps for rapid cycling *Brassica rapa* and *B. oleracea*. Theor Appl Genet 120(1):31–43. https://doi.org/10.1007/s00122-009-1157-4

Izzah NK, Lee J, Jayakodi M, Perumal S, Jin M, Park BS, Ahn K, Yang TJ (2014) Transcriptome sequencing of two parental lines of cabbage (*Brassica oleracea* L. var. capitata L.) and construction of an EST-based genetic map. BMC Genomics 15:149. https://doi.org/10.1186/1471-2164-15-149

Kaneko Y, Bang SW (2014) Interspecific and intergeneric hybridization and chromosomal engineering of

Brassicaceae crops. Breed Sci 64(1):14–22. https://doi.org/10.1270/jsbbs.64.14

Kato T, Hatakeyama K, Fukino N, Matsumoto S (2012) Identificaiton of a clubroot resistance locus conferring resistance to a *Plasmodiophora brassicae* classified into pathotype group 3 in Chinese cabbage (*Brassica rapa* L.). Breed Sci 62(3):282–287. https://doi.org/10.1270/jsbbs.62.282

Kearsey MJ, Ramsay LD, Jennings DE, Lydiate DJ, Bohuon EJ, Marshall DF (1996) Higher recombination frequencies in female compared to male meisoses in *Brassica oleracea*. Theor Appl Genet 92(3–4):363–367. https://doi.org/10.1007/bf00223680

Kianian SF, Quiros CF (1992) Generation of a *Brassica oleracea* composite RFLP map: linkage arrangements among various populations and evolutionary implications. Theor Appl Genet 84(5–6):544–554. https://doi.org/10.1007/BF00224150

Kifuji Y, Hanzawa H, Terasawa Y, Ashutosh Nishio T (2012) QTL analysis of black rot resistance in cabbage using newly developed EST-SNP markers. Euphytica 190(2):289–295. https://doi.org/10.1007/s10681-012-0847-1

Lagercrantz U (1998) Comparative mapping between *Arabidopsis thaliana* and *Brassica nigra* indicates that Brassica genomes have evolved through extensive genome replication accompanied by chromosome fusions and frequent rearrangements. Genetics 150 (3):1217–1228

Lan TH, Paterson AH (2000) Comparative mapping of quantitative trait loci sculpting the curd of *Brassica oleracea*. Genetics 155(4):1927–1954

Landry BS, Hubert N, Crete R, Chang MS, Lincoln SE, Etoh T (1992) A genetic map for *Brassica oleracea* based on RFLP markers detected with expressed DNA sequences and mapping of resistance genes to race 2 of *Plasmodiophora brassicae* (Woronin). Genome 35 (3):409–420. https://doi.org/10.1139/g92-061

Lee J, Izzah NK, Choi B-S, Joh HJ, Lee S-C, Perumal S, Seo J, Ahn K, Jo EJ, Choi GJ, Nou I-S, Yu Y, Yang T-J (2015a) Genotyping-by-sequencing map permits identification of clubroot resistance QTLs and revision of the reference genome assembly in cabbage (*Brassica oleracea* L.). DNA Res. https://doi.org/10.1093/dnares/dsv034

Lee J, Izzah NK, Jayakodi M, Perumal S, Joh HJ, Lee HJ, Lee S-C, Park JY, Yang K-W, Nou I-S, Seo J, Yoo J, Suh Y, Ahn K, Lee JH, Choi GJ, Yu Y, Kim H, Yang T-J (2015b) Genome-wide SNP identification and QTL mapping for black rot resistance in cabbage. BMC Plant Biol 15(1):1–11. https://doi.org/10.1186/s12870-015-0424-6

Li G, Gao M, Yang B, Quiros CF (2003a) Gene for gene alignment between the *Brassica* and Arabidopsis genomes by direct transcriptome mapping. Theor Appl Genet 107:168–180. https://doi.org/10.1007/s00122-003-1236-x

Li G, Quiros CF (2002) Genetic analysis, expression and molecular characterization of *Bo-GSL-ELONG*, a major gene involved in the aliphatic glucosinolate pathway of *Brassica* species. Genetics 162:1937–1943

Li G, Quiros CF (2003) In planta side-chain glucosinolate modification in Arabidopsis by introduction of dioxygenase *Brassica* homolog *BoGSL-ALK*. Theor Appl Genet 106:1116–1121. https://doi.org/10.1007/s00122-002-1161-4

Li G, Riaz A, Goyal S, Abel S, Quiros CF (2001a) Inheritance of three major genes involved in the synthesis of aliphatic glucosinolates in *Brassica oleracea*. J Amer Soc Hort Sci 126(4):427–431

Li L, Lu S, O'Halloran DM, Garvin DF, Vrebalov J (2003b) High-resolution genetic and physical mapping of the cauliflower high-beta-carotene gene *Or* (*Orange*). Mol Genet Genom 270(2):132–138. https://doi.org/10.1007/s00438-003-0904-5

Li L, Paolillo DJ, Parthasarathy MV, Dimuzio EM, Garvin DF (2001b) A novel gene mutation that confers abnormal patterns of beta-carotene accumulation in cauliflower (*Brassica oleracea* var. *botrytis*). Plant J 26 (1):59–67

Lin HJ, Probst-Hensch NM, Louie AD, Kau IH, Witte JS, Ingles SA, Frankl HD, Lee ER, Haile RW (1998) Glutathione transferase null genotype, broccoli, and lower prevalence of colorectal adenomas. Cancer Epidemiol Biomarkers Prev 7(8):647–652

Liu S, Liu Y, Yang X, Tong C, Edwards D, Parkin IAP, Zhao M, Ma J, Yu J, Huang S, Wang X, Wang J, Lu K, Fang Z, Bancroft I, Yang T-J, Hu Q, Wang X, Yue Z, Li H, Yang L, Wu J, Zhou Q, Wang W, King GJ, Pires JC, Lu C, Wu Z, Sampath P, Wang Z, Guo H, Pan S, Yang L, Min J, Zhang D, Jin D, Li W, Belcram H, Tu J, Guan M, Qi C, Du D, Li J, Jiang L, Batley J, Sharpe AG, Park B-S, Ruperao P, Cheng F, Waminal NE, Huang Y, Dong C, Wang L, Li J, Hu Z, Zhuang M, Huang Y, Huang J, Shi J, Mei D, Liu J, Lee T-H, Wang J, Jin H, Li Z, Li X, Zhang J, Xiao L, Zhou Y, Liu Z, Liu X, Qin R, Tang X, Liu W, Wang Y, Zhang Y, Lee J, Kim HH, Denoeud F, Xu X, Liang X, Hua W, Wang X, Wang J, Chalhoub B, Paterson AH (2014) The *Brassica oleracea* genome reveals the asymmetrical evolution of polyploid genomes. Nat Commun 5

Lowe aJ, Moule C, Trick M, Edwards KJ (2004) Efficient large scale development of microsatellites for marker and mapping applications in *Brassica* crop species. Theor Appl Genet 108:1103–1112. doi:10.1007/s00122-003-1522-7

Lu S, Van Eck J, Zhou X, Lopez AB, O'Halloran DM, Cosman KM, Conlin BJ, Paolillo DJ, Garvin DF, Vrebalov J, Kochian LV, Kupper H, Earle ED, Cao J, Li L (2006) The cauliflower *Or* gene encodes a DnaJ cysteine-rich domain-containing protein that mediates high levels of beta-carotene accumulation. Plant Cell 18(12):3594–3605. https://doi.org/10.1105/tpc.106.046417

Mei J, Ding Y, Lu K, Wei D, Liu Y, Disi JO, Li J, Liu L, Liu S, McKay J, Qian W (2012) Identification of genomic regions involved in resistance against *Sclerotinia sclerotiorum* from wild *Brassica oleracea*. Theor Appl Genet 126(2):549–556. https://doi.org/10.1007/s00122-012-2000-x

Mei J, Ding Y, Lu K, Wei D, Liu Y, Disi JO, Li J, Liu L, Liu S, McKay J, Qian W (2013) Identification of genomic regions involved in resistance against *Sclerotinia sclerotiorum* from wild *Brassica oleracea*. Theor Appl Genet 126(2):549–556. https://doi.org/10.1007/s00122-012-2000-x

Mei J, Qian L, Disi JO, Yang X, Li Q, Li J, Frauen M, Cai D, Qian W (2010) Identification of resistant sources against *Sclerotinia sclerotiorum* in *Brassica* species with emphasis on *B. oleracea*. Euphytica 177 (3):393–399. https://doi.org/10.1007/s10681-010-0274-0

Moriguchi K, Kimizuka-Takagi C, Ishii K, Nomura K (1999) A Genetic map based on RAPD, RFLP, Isozyme, morphological markers and QTL analysis for clubroot resistance in *Brassica oleracea*. Breed Sci 49(4):257–265. https://doi.org/10.1270/jsbbs.49.257

Nagaoka T, Doullah MA, Matsumoto S, Kawasaki S, Ishikawa T, Hori H, Okazaki K (2010) Identification of QTLs that control clubroot resistance in *Brassica oleracea* and comparative analysis of clubroot resistance genes between *B. rapa* and *B. oleracea*. Theor Appl Genet 120(7):1335–1346. https://doi.org/10.1007/s00122-010-1259-z

Ogura H (1968) Studies on the new male-sterility in Japanese radish, with special reference to the utilization of this sterility towards the practical raising of hybrid seeds. Mem Fac Agri Kagoshima Univ 6:39–78

Okazaki K, Sakamoto K, Kikuchi R, Saito A, Togashi E, Kuginuki Y, Matsumoto S, Hirai M (2007) Mapping and characterization of FLC homologs and QTL analysis of flowering time in *Brassica oleracea*. Theor Appl Genet 114(4):595–608. https://doi.org/10.1007/s00122-006-0460-6

Osborn TC, Kole C, Parkin IA, Sharpe AG, Kuiper M, Lydiate DJ, Trick M (1997) Comparison of flowering time genes in *Brassica rapa*, *B. napus* and *Arabidopsis thaliana*. Genetics 146(3):1123–1129

Piao ZY, Deng YQ, Choi SR, Park YJ, Lim YP (2004) SCAR and CAPS mapping of *CRb*, a gene conferring resistance to *Plasmodiophora brassicae* in Chinese cabbage (*Brassica rapa* ssp. *pekinensis*). Theor Appl Genet 108(8):1458–1465. https://doi.org/10.1007/s00122-003-1577-5

Piao ZY, Ramchiary N, Lim YP (2009) Genetics of clubroot resistance in *Brassica* species. J Plant Growth Regul 28(3):252–264. https://doi.org/10.1007/s00344-009-9093-8

Piquemal J, Cinquin E, Couton F, Rondeau C, Seignoret E, Doucet I, Perret D, Villeger M-J, Vincourt P, Blanchard P (2005) Construction of an oilseed rape (*Brassica napus* L.) genetic map with SSR markers. Theor Appl Genet 111:1514–1523. https://doi.org/10.1007/s00122-005-0080-6

Podsędek A (2007) Natural antioxidants and antioxidant capacity of Brassica vegetables: A review. LWT Food Sci Technol 40(1):1–11. https://doi.org/10.1016/j.lwt.2005.07.023

Qiu D, Gao M, Li G, Quiros C (2009) Comparative sequence analysis for *Brassica oleracea* with similar sequences in *B. rapa* and *Arabidopsis thaliana*. Plant Cell Rep 28:649–661. https://doi.org/10.1007/s00299-008-0661-3

Quiros CF (2000) DNA-based marker maps of *Brassica*. In: Phillips RL, Vasil JK (eds) DNA-based markers in plants. Kluwer, Dordrecht, pp 201–238

Quiros CF, Paterson AH (2004) Genome mapping and analysis in *Brassica*. In: Pua EC, Douglas CJ (eds) Biotechnology in agriculture and forestry, vol 54. Springer, New York, pp 31–42

Raman H, Raman R, Larkan N (2013) Genetic dissection of blackleg resistance loci in rapeseed (*Brassica napus* L.). In: Andersen SB (ed) Plant breeding from laboratories to fields. InTech. https://doi.org/10.5772/53611

Ramsay LD, Jennings DE, Kearsey MJ, Marshall DF, Bohuon EJ, Arthur AE, Lydiate DJ (1996) The construction of a substitution library of recombinant backcross lines in *Brassica oleracea* for the precision mapping of quantitative trait loci. Genome 39(3):558–567

Ren J, Liu Z, Niu R, Feng H (2015) Mapping of *Re*, a gene conferring the red leaf trait in ornamental kale (*Brassica oleracea* L. var. acephala). Plant Breed 134 (4):494–500. https://doi.org/10.1111/pbr.12286

Renard M, Delourme R, Mesquida J, Pelletier G, Primard C, Boulidard L, Dore C, Ruffio V, Herve Y, Morice J (1992) Male sterilities and F_1 hybrids in *Brassica*. In: Dattee Y, Dumas C, Gallais A, Mason S (eds) Reproductive biology and plant breeding. Springer, Berlin, Heidelberg, New York, pp 107–119

Rocherieux J, Glory P, Giboulot A, Boury S, Barbeyron G, Thomas G, Manzanares-Dauleux MJ (2004) Isolate-specific and broad-spectrum QTLs are involved in the control of clubroot in *Brassica oleracea*. Theor Appl Genet 108(8):1555–1563. https://doi.org/10.1007/s00122-003-1580-x

Sakamoto K, Saito A, Hayashida N, Taguchi G, Matsumoto E (2008) Mapping of isolate-specific QTLs for clubroot resistance in Chinese cabbage (*Brassica rapa* L. ssp. *pekinensis*). Theor Appl Genet 117(5):759–767. https://doi.org/10.1007/s00122-008-0817-0

Salmon A, Clotault J, Jenczewski E, Chable V, Manzanares-Dauleux MJ (2008) *Brassica oleracea* displays a high level of DNA methylation polymorphism. Plant Sci 174(1):61–70. https://doi.org/10.1016/j.plantsci.2007.09.012

Schranz ME, Quijada P, Sung SB, Lukens L, Amasino R, Osborn TC (2002) Characterization and effects of the

replicated flowering time gene FLC in *Brassica rapa*. Genetics 162(3):1457–1468
Sebastian RL, Howell EC, King GJ, Marshall DF, Kearsey MJ (2000) An integrated AFLP and RFLP *Brassica oleracea* linkage map from two morphologically distinct doubled-haploid mapping populations. Theor Appl Genet 100(1):75–81. https://doi.org/10.1007/s001220050011
Shinada T, Kikuchi Y, Fujimoto R, Kishitani S (2006) An alloplasmic male-sterile line of *Brassica oleracea* harboring the mitochondria from *Diplotaxis muralis* expresses a novel chimeric open reading frame, *orf72*. Plant Cell Physiol 47(4):549–553. https://doi.org/10.1093/pcp/pcj014
Singh BK, Sharma SR, Singh B (2011a) Genetic combining ability for concentration of mineral elements in cabbage head (*Brassica oleracea* var. *capitata* L.). Euphytica 184(2):265–273. https://doi.org/10.1007/s10681-011-0598-4
Singh D, Dhar S, Yadava DK (2011b) Genetic and pathogenic variability of Indian strains of *Xanthomonas campestris* pv. *campestris* causing black rot disease in crucifers. Curr Microbiol 63 (6):551–560. https://doi.org/10.1007/s00284-011-0024-0
Slocum MK, Figdore SS, Kennard WC, Suzuki JY, Osborn TC (1990) Linkage arrangement of restriction fragment length polymorphism loci in *Brassica oleracea*. Theor Appl Genet 80(1):57–64. https://doi.org/10.1007/BF00224016
Sotelo T, Cartea ME, Velasco P, Soengas P (2014a) Identification of antioxidant capacity related QTLs in *Brassica oleracea*. PLoS One 9(9):e107290. https://doi.org/10.1371/journal.pone.0107290
Sotelo T, Soengas P, Velasco P, Rodríguez VM, Cartea ME (2014b) Identification of metabolic QTLs and candidate genes for glucosinolate synthesis in *Brassica oleracea* leaves, seeds and flower buds. PLoS One 9(3):e91428. https://doi.org/10.1371/journal.pone.0091428
Srivastava M, Akhoon BA, Gupta SK, Gupta SK (2010) Development of resistance against blackleg disease in *Brassica oleracea* var. *botrytis* through in silico methods. Fungal Genet Biol 47(10):800–808. http://dx.doi.org/10.1016/j.fgb.2010.06.014
Suwabe K, Tsukazaki H, Iketani H, Hatakeyama K, Kondo M, Fujimura M, Nunome T, Fukuoka H, Hirai M, Matsumoto S (2006) Simple sequence repeat based comparative genomics between *Brassica rapa* and *Arabidopsis thaliana:* the genetic origin of clubroot resistance. Genetics 173:309–319. https://doi.org/10.1534/genetics.104.038968
Tadege M, Sheldon CC, Helliwell CA, Stoutjesdijk P, Dennis ES, Peacock WJ (2001) Control of flowering time by FLC orthologues in *Brassica napus*. Plant J 28 (5):545–553. https://doi.org/10.1046/j.1365-313X.2001.01182.x
Tanaka N, Niikura S (2006) Genetic analysis of the developmental characteristics related to the earliness of head formation in cabbage (*Brassica oleracea* L.). Breed Sci 56(2):147–153. https://doi.org/10.1270/jsbbs.56.147
Tonu NN, Doullah MA, Shimizu M, Karim M, Kawanabe T, Fujimoto R, Okazaki K (2013) Comparison of positions of QTLs conferring resistance to *Xanthomonas campestris* pv. *campestris* in *Brassica oleracea*. Amer J Plant Sci 4:11–20
Voorrips R (1995) *Plasmodiophora brassicae:* aspects of pathogenesis and resistance in *Brassica oleracea*. Euphytica 83(2):139–146. https://doi.org/10.1007/BF01678041
Voorrips RE, Jongerius MC, Kanne HJ (1997) Mapping of two genes for resistance to clubroot (*Plasmodiophora brassicae*) in a population of doubled haploid lines of *Brassica oleracea* by means of RFLP and AFLP markers. Theor Appl Genet 94(1):75–82. https://doi.org/10.1007/s001220050384
Voorrips RE, Kanne HJ (1997) Genetic analysis of resistance to clubroot (*Plasmodiophora brassicae*) in *Brassica oleracea*. II. Quantitative analysis of root symptom measurements. Euphytica 93(1):41–48. https://doi.org/10.1023/a:1002970103683
Voorrips RE, Visser DL (1993) Examination of resistance to clubroot in accessions of *Brassica oleracea* using a glasshouse seedling test. Neth J Plant Pathol 99(5–6):269–276. https://doi.org/10.1007/Bf01974308
Walley PG, Carder J, Skipper E, Mathas E, Lynn J, Pink D, Buchanan-Wollaston V (2012) A new broccoli x broccoli immortal mapping population and framework genetic map: tools for breeders and complex trait analysis. Theor Appl Genet 124 (3):467–484. https://doi.org/10.1007/s00122-011-1721-6
Wang W, Huang S, Liu Y, Fang Z, Yang L, Hua W, Yuan S, Liu S, Sun J, Zhuang M, Zhang Y, Zeng A (2012) Construction and analysis of a high-density genetic linkage map in cabbage (*Brassica oleracea* L. var. *capitata*). BMC Genomics 13:523. https://doi.org/10.1186/1471-2164-13-523
Wang YS, Liu ZY, Li YF, Zhang Y, Yang XF, Feng H (2013) Identification of sequence-related amplified polymorphism markers linked to the red leaf trait in ornamental kale (*Brassica oleracea* L. var. *acephala*). Genet Mol Res 12(2):870–877. https://doi.org/10.4238/2013.april.2.3
Williams PH (1980) Black rot: a continuing threat to world crucifers. Plant Dis 64:736–742
Yang B, Quiros CF (2010) Survey of glucosinolate variation in leaves of *Brassica rapa* crops. Genet Resour Crop Evol 57(7):1079–1089. https://doi.org/10.1007/s10722-010-9549-5
Zhang B, Hu Z, Zhang Y, Li Y, Zhou S, Chen G (2011) A putative functional MYB transcription factor induced by low temperature regulates anthocyanin biosynthesis in purple kale (*Brassica Oleracea* var. *acephala* f. tricolor). Plant Cell Rep 31(2):281–289. https://doi.org/10.1007/s00299-011-1162-3

Zhang J, Li H, Zhang M, Hui M, Wang Q, Li L, Zhang L (2013) Fine mapping and identification of candidate *Br-or* gene controlling orange head of Chinese cabbage (*Brassica rapa* L. ssp. *pekinensis*). Mol Breed 32(4):799–805. https://doi.org/10.1007/s11032-013-9907-z

Zhu P, Cheng M, Feng X, Xiong Y, Liu C, Kang Y (2015) Mapping of *Pi*, a gene conferring pink leaf in ornamental kale (*Brassica oleracea* L. var. *acephala* DC). Euphytica 207(2):377–385. https://doi.org/10.1007/s10681-015-1555-4

A Physical Map of the *Brassica oleracea* Genome

3

Xiyin Wang

Abstract

After the shared polyploidization events affecting *Arabidopsis thaliana*, Brassica plants have been further affected by extra events, resulting in complex genome structure and increasing gene copy numbers and thereof complexity of regulatory networks. Here, by using high-information-content fingerprinting (HICF) of bacterial artificial chromosome (BAC) clones, we constructed a physical map of the *Brassica oleracea* genome, and the map was used to assemble the whole-genome sequence of *B. olearacea*. In total, 2,907 contigs of two or more BACs were produced. By mapping the BAC-end sequences onto sequenced genomes, we explored genome-wide microsynteny between Arabidopsis and Brassica plants (*B. oleracea* and *B. rapa*). Besides, by using the map and its BAC information, we performed comparative analysis between *B. oleracea* with Arabidopsis and several other available eudicot genomes. This effort helped find evolution of different genomic regions, such as euchromatin and heterochromatin, and showed appreciable shadowing effect produced by more ancient polyploidization events. We found that these recursive polyploidization events resulted in interweaving patterns among contigs much increasing genomic complexity. In the past, without the information of whole-genome sequence, the analysis using physical map provided an approach to understand the structure and evolution of complex genomes.

3.1 Introduction

3.1.1 Extensive Polyploidization Contributed to Plant Genome Formation

More and more evidence showed that the extant plants, often having a small genome sizes and small numbers of chromosomes, have been recurrently affected by polyploidization events (Blanc et al. 2003; Bowers et al. 2003; Tang et al. 2008; Soltis et al. 2009). These events have resulted in a high level of duplicated DNA regions and functional genes. The existence of duplicated regions, especially younger ones produced by recent polyploidies, constitutes an obstacle to produce a reliable and credible

X. Wang (✉)
Plant Genome Mapping Laboratory, University of Georgia, Athens, GA 30602, USA
e-mail: wangxiyin@vip.sina.com

National Key Laboratory for North China Crop Improvement and Regulation, Agriculture University of Hebei, Baoding, China

S. Liu et al. (eds.), *The Brassica oleracea Genome*, Compendium of Plant Genomes,
https://doi.org/10.1007/978-3-030-31005-9_3

genome sequence. Besides, the existence of enormous repetitive sequences in many extant plant genomes constitutes another obstacle. Therefore, before using the third-generation sequencing technologies, many plant genomes were sequenced with draft or unfinished genome sequences, and failed to support inter-specific comparison of genome organization and structure. These comparisons are indispensable to understand the evolution, speciation, and functional innovation of plants. In concert with draft or complete sequence assemblies, independent high-resolution physical maps will facilitate the correct ordering of DNA segments on chromosomes. Thus the production and usage of physical map will greatly improve the precise details of changes in genome organization revealed by multiple species comparison, such as chromosomal breaks, fusions and divergent and/or complementary evolution between syntenic DNA segments (Lewin et al. 2009).

3.1.2 Sequencing Brassica Genomes

The genus *Brassica* provide many daily foods and vegetables, thereof crucial economic importance of human life (Paterson et al. 2001). Actually, crops of the genus *Brassica* are extensively used as food in the cuisines, as fodder and forage, and provide much of edible vegetable oil supplies. Together with *B. rapa* and the allotetraploid descendent (*B. napus*), *B. oleracea* comprise many vegetables. The extreme morphological characteristics of *B. oleracea* plants make them to form an especially attractive system, to elucidate biological and regulatory processes. Besides, their relatively small genomes also facilitate them to be model plants for genome structure and evolution studies. It was predicted that *B. oleracea* has a genome size of ~696 Mb DNA (Johnston et al. 2005), which was sequenced to add knowledge of its biology (Liu et al. 2014). The genomes of *B. rapa* and *B. napus* were also available (Wang et al. 2011; Chalhoub et al. 2014), laying solid foundation for Brassica scientific and agricultural research.

Due to their close relationship with the plant model species *Arabidopsis thaliana*, which genome is the firstly sequenced (The Arabidopsis Genome Initiative 2000), it was anticipated that *Brassica* crop improvement would be straightforward by transferential using the knowledge from *A. thaliana*. However, although the physiology and regulatory biology of Arabidopsis and Brassica are thought to be very similar, the genomes of *Brassica* species are well known to be much more complex than that of *A. thaliana* (Parkin et al. 2003; Town et al. 2006), as a result of multiple and extra rounds of polyploidy contributing to the formation of each lineage of Brassica plants(Lysak et al. 2005; Schranz et al. 2006). In fact, one round of ancient hexaploidization or whole-genome triplication (gamma) had produced the common ancestor of core eudicots. Millions of years later, two tetraploidization events or whole-genome duplications (beta and alpha) had occurred before the Arabidopsis-Brassica split, and they were demonstrated to have occurred in the Brassicaceae species, after their divergence from *Carica papaya* (Jaillon et al. 2007; Ming et al. 2008; Tang et al. 2008). The triplicated Brassica genome segments diverged from a common ancestor soon after the Arabidopsis-Brassica split implied that this recent hexaploidization or genome triplication contributed to speciation of 240 *Brassica* plants (Town et al. 2006). The genetic maps developed for Brassica plants facilitated the identification of a mixture of single, duplicated, and triplicated Brassica chromosomal segments relative to Arabidopsis (Lan et al. 2000; Schmidt et al. 2001; Babula et al. 2003; Lukens et al. 2003; Parkin et al. 2003). Using a cytogenetic approach, it was also shown that the whole-genome triplication affected the chromosome numbers and structures (Lysak et al. 2005).

Following polyploidization, a process of diploidization that recovers diploid status ensues, characterized by extensive gene loss, transposon activation, and chromosomal rearrangements

(Paterson et al. 2004; Wang et al. 2005). Diploidization should have occurred especially after autoploidization, which resulted in multiple homologous chromosomes executing complex pairing. It could also occur after allotetraploidization, often involving homeologous chromosomes, which may occasionally execute multiple pairing among certain regions of the homologous and homeologous chromosomes. Actually, it seems that diploid heredity may have some advantage over polyploid heredity, due at least partially to the possibility of illegitimate recombination between homeologous chromosomes after polyploidization (Wang et al. 2009), and illegitimate recombination may lead to genome instability, gene conversion, and gene dosage disorder. In a diploid species, homeologous recombination would be nearly completely restricted. Nonetheless, it has been repeatedly stated that polyploidy may have been a prominent factor in the evolution of genetic novelties which led to the global success of land plants (Wendel 2000; De Bodt et al. 2005; Soltis et al. 2009). Anyway, this diploidization process could further complicated knowledge transfer from the model plant Arabidopsis to Brassica.

Grossly, all these make *Brassica* genus a good plant system to study genomic and chromosomal changes associated with recursive polyploidization. Here, to support the sequencing of Brassica 'C' genome, we produced high-information-content fingerprinting (HICF) of bacterial artificial chromosome (BAC) clones, and with them constructed a physical map of *B. oleracea.* Then, by integratively using *B. rapa* physical map, we managed to show genome-wide and inter-Arabidopsis-Brassica microsynteny, and find probable (peri)centromere-related contigs. Third, by mapping the *B. oleracea* map onto Arabidopsis and the other available core eudicot genomes helped us find appreciable shadowing effect related to more ancient polyploidy. This effort showed interweaving patterns among constructed contigs, and these interweaving patterns vividly showed the genomic complexity, being a potential obstacle to construct whole-genome sequence assembly and perform future comparative genomics analysis. Honestly, the later availability genome sequences of the *B. Olerecea* and the other Brassica plants provided much complete information of their genomes. However, these analyses with a physical map are still interesting in showing us what it can tell us about genome evolution.

3.2 Methodologies for Physical Mapping

3.2.1 Probe Design and Hybridization

In total, 4226 *B. oleracea* overgo probes, 40 bp each, were successfully hybridized to the BOTO1 BAC clone library. Using the same method as previously described, we designed these probes from Arabidopsis gene sequences and hybridized them to the libraries (Bowers et al. 2005). Briefly, by using BLASTN, source sequences were searched against known plant sequences to find conserved domains, and then compared to known plant repeats to screen out possible repeats. The selected sequences were then cut into 40 bp segments, and then screened for Guanine + Cytosine content between 40 and 60%. We further selected a subset of 2882 probes that were successfully hybridized onto the BACs and integrated in the assembly process.

3.2.2 BAC Fingerprinting

The BAC library BOTO1 used here was obtained from partial *Hind*III digest of *Brassica oleracea* DNA. We adopted the high-information-content fingerprinting (HICF) approach for its well-established format with a SNAPshot labeling toolkit. The enzymes, *Eco*RI, *Bam*HI, *Xba*I, *Xho*I and *Hha*I, were used to perform the digestion. After the first four enzyme cuts, the ends of each fragment were differentially labeled using fluorochrome tagged ddNTPs, and the last enzyme was used to further reduce fragment size and produce a blunt end. An ABI sequencer was used to generate the fingerprints and GeneMapper software v4.0 was used to generate the size files after processing the chromatograms. Only the fragments with 50–600 bp were involved for further analysis, in that fragments beyond this range were proved unreliable.

We run FPC (v9.0) to construct the contigs, and the fragments (bands) were processed to produce FPC-readable data set (Soderlund et al. 2000). Because FPC does not use color labels or fractional sizes, therefore the size of each fragment was multiplied by 10; and then the decimal part was dropped, which will not lose significant information. Actually, the above processes produced fragments with size 500–6,000 bp. Then, the color labels were converted to non-overlapping numeric ranges by adding offset values 6,000, 12,000, or 18,000 to three of the four colors, which eventually resulted in fragments ranging from 0 to 25,000.

3.2.3 Physical Map Assembly

As to our experience, neighboring well contamination could produce the most problem. To avoid this possibility, before constructing contigs, we removed likely contaminated BACs in the dataset by using a de-contamination process implemented in FPC software. We performed an initial round of contig construction with parameters: cutoff = 1e-50 and tolerance = 4 and best of 100 repetitive constructions. Then, a quality checking process was performed to eliminate Q-clones for contigs involving more than 15% possible Q-clones. Several rounds of end-to-end merges and singleton-to-contig merges were then performed with less and less stringent parameter settings (Fig. 3.1).

Probes were labeled using P-32 and then applied to macroarrays (18,432 BACs per membrane). Multiplex experiments are made by involving 576 probes in each experiment, in pools of 24 probes per bottle, by rows, columns, and diagonals of a 24 × 24 array of probes. Films were then manually checked and scored, and scores were digitized using ABBYY FINEREADER, a text-recognition software. Data were eventually deconvoluted and stored in MS Access database system 'BACMan' developed by ourselves.

3.2.4 BAC-End Sequencing and Analysis

As described previously (Thon et al. 2004), a subset of BACs were sequenced with their ends, producing 85,317 BAC-end sequences (https://www.ncbi.nlm.nih.gov/). By searching against Brassica repetitive sequence database in the Institute for Genome Research (TIGR), our own extended Brassica repeats database, especially two Brassica-centromere-specific repetitive sequences (Lim et al. 2007), the BAC-end sequences were related to be repeats, and specifically, centromere repeats.

3.2.5 Inferring RNA-, Chloroplast- and Mitochondrion-Related Contigs

We downloaded from GenBank the following sequences, involving the eudicot RNA sequences, *A. thaliana* chloroplast sequence

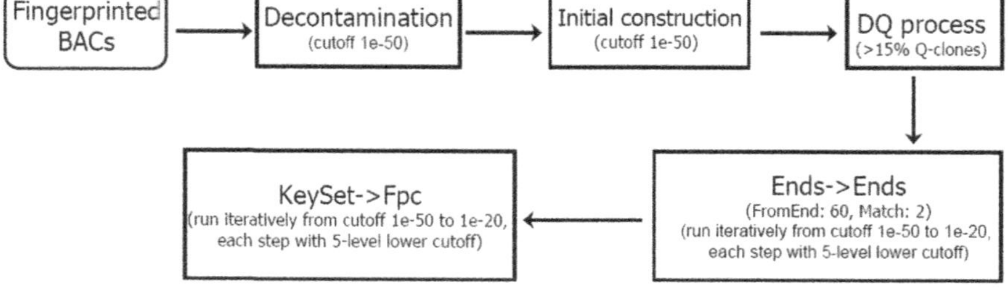

Fig. 3.1 Streamline of running the software FPC

(AP000423.1), and *B. napus* mitochondrion sequence (AP006444.1), and against them we searched *B. olerecea* BAC-end sequences at BLASTN E-value <1e-5. If >20% of BAC-end sequences of a contig matched these specific sequences, the contig was related to RNA-, chloroplast- and/or mitochondrion, respectively.

3.2.6 Comparative Genomics Analysis

To perform comparative analysis, by BLASTN-searching BAC-end sequences and probe sequences against eudicot sequences (E-value <1e-10 for Arabidopsis and E-value <1e-5 for other eudicots), the constructed contigs were anchored to Arabidopsis (The Arabidopsis Genome Initiative 2000), papapa (Ming et al. 2008), poplar (Tuskan et al. 2006), and grape (Jaillon et al. 2007) genomes. BAC-end sequences and probe sequences having more than 50 hits were dropped from the following analysis. DNA syntenic regions between species were identified by linking neighboring hits 200 Kb, and therefore we checked whether a *B. oleracea* contig could be related to Arabidopsis duplicated regions produced by recursive polyploidization (Bowers et al. 2003).

3.2.7 Comparative Analysis of the *Brassica rapa* Physical Map

To increase the credibility of the analysis, the *B. rapa* contigs previously published (Mun et al. 2008) were involved and anchored onto the Arabidopsis genome by using its 100,666 BAC-end sequences (https://www.ncbi.nlm.nih.gov/).

3.3 Outcomes of Physical Mapping

3.3.1 Fingerprinting BAC Clones

A total of 61,871 clones were fingerprinted on 192,384-well plates. Clones <30 and >250 bands were removed from the data provided for earlier FPC analysis. Actually, on doing FPC analysis, the clones with >200 bands were further removed, producing a dataset including 53,048 clones.

With certain subplates, we repeatedly performed fingerprinting to ensure the quality, with 72 subplates or 5,184 BACs duplicated, and 12 subplates or 830 BACs triplicated. At FPC cutoff 1e-50, we found that the replicated BACs had an opportunity 72.1% being overlapped on the same contig. This showed an appreciable repeatability and credibility of the contigs to assemble. For a BAC having repeats, the fingerprinting output having numbers of bands $\sim$120, the global average, was used in following assembling process.

At E-value 1e-5, a relatively loose criteria, 35,064 BAC-end sequences out of 85,317 matched Brassica repetitive sequences, indicating $\sim$40% of clones to be repeat-related. Considering that all BACs would help produce a comprehensive assembly characterizing the repetitive regions, we preserved all of them for further contig assembling process rather than removing those likely repeat-related ones.

3.3.2 Assembling Contigs

In consideration of well-to-well contamination, adopting a novel de-contamination function provided in FPC v9.0, we screened likely contaminated wells before assembling. If a clone in a well has many overlapping bands to a significance level (e.g. cutoff 1e-50), with one of its neighboring 7 $\times$ 7 square of wells, it was considered to have been likely contaminated. Then, 5,477 clones were found to have been likely contaminated and therefore removed from assembling process.

Grossly, 46,006 BACs were selected to assemble contigs, producing 2907 contigs, each containing $\geq$2 BACs, and 2,323 singletons. Averagely, a contig involved 8.7 BACs and 3.2 overgo probes. There were two contigs (ctg03293 and ctg02560) involving >1000 BACs, $\sim$60% of whose BAC-end sequences were repeat-related. There were five contigs involving <100 but <1000 BACs. Sixteen contigs are RNA-related, likely being helpful to decipher the rRNA and rRNA genes. Six contigs were found to be likely chloroplast-related, and five contigs were likely mitochondrion-related, with two contigs being both chloroplast- and mitochondrion-related, showing possible organellar gene transfer, as reported previously in Arabidopsis (Martin 2003), and sorghum (Bowers et al. 2005).

3.3.3 Mapping onto Arabidopsis Genome

With the help of BAC-end sequences and probe sequences, both *B. oleracea* and *B. rapa* contigs were mapped onto Arabidopsis genome sequences. At the extension parameter 200 Kb, that is, the neighboring hits $\leq$200 Kb were used to infer DNA synteny between *B. oleracea* and Arabidopsis, we inferred the longest syntenic regions with ~870 Kb, and with most regions <400 Kb (Fig. 3.2a). We also anchored *B. rapa* contigs onto Arabidopsis sequences, while the extension parameter was reduced to 100 Kb in that substantially more BESs made it easier to find cross-species synteny. Grossly, 1,990 *B. oleracea* and 1,006 *B. rapa* contigs (~70% of the total, respectively) hit $\geq$1 Arabidopsis regions. Cross-species DNA sequence similarity made a peak at 92% (Fig. 3.2b), supporting Arabidopsis-Brassica divergence around 14.5–20.4 million years (Blanc et al. 2003; Bowers et al. 2003). Notably, Blast E-value showed a bi-modal distribution (Fig. 3.2c), reflecting the ancient polyploidization events.

There found evidence that polyploidization events have resulted a homology network in the extant Brassica genomes. *B. oleracea* and *B. rapa* contigs covered ~88 and 93% of Arabidopsis genome sequences, respectively (Figs. 3.3, 3.4 and 3.5), with >70% regions covered in a depth $\geq$2. This could be explained by the existence of multiple homeologous regions in Brassica genomes. The peak is about 3, covering ~1/5–1/4 of Arabidopsis genome sequences. Clearly, there is a sharp decrease from coverage 3 to 4, supporting a triplication event in Brassica common ancestor after its divergence from Arabidopsis lineage. Notably, Brassica contigs covered 20% of Arabidopsis genome to the depths $\geq$4, well explained by the shadow effect resulted from more ancient genome duplications.

Therefore, we checked Arabidopsis homeologous regions previously related to polyploidization events, and found that 9.3% (or 186) of all anchored *B. oleracea* contigs were anchored to the more recent polyploidization events to shape Arabidopsis genome (namely, alpha tetraploidization), and another 2.7% (or 54) related to two more ancient events (namely, beta tetraploidization and gamma hexaploidization). We found that it is possible to separate orthologous from outparalogous regions (or regions produced by genomic duplications before Arabidopsis-Brassica split). The orthologous regions were found to often have E-values <1e-30. Comparatively, the outparalogous ones were found to have E-value ~1e-10.

3.3.4 Regions Related to Heterochromatin or Euchromatin

We found well preserved Arabidopsis-Brassica synteny, virtually mainly in the euchromatin but not the heterochromatin (often pericentromeric) regions (Fig. 3.3). The regions related to different types of chromatins were predicted according to densities of gene and repetitive sequences. Actually, ~14% of Arabidopsis sequences were not covered by *B. oleracea* contigs, which were mainly related to the Arabidopsis pericentromeric regions (Fig. 3.3). In comparison, among 1,990 *B. oleracea* contigs (note: excluding the largest 5 ones, being likely badly constructed mosaic), 97% (or 1920) were aligned to the 104 Mb euchromatic regions in Arabidopsis, involving 80% (or 2,316) anchored probes and 91% (or 32,415) anchored BACs, whereas, only 6.7% contigs, 20% probes, and 9% BACs were aligned to the 15 Mb heterochromatic regions. About 3% *B. oleracea* contigs could be anchored to both types of chromatin regions.

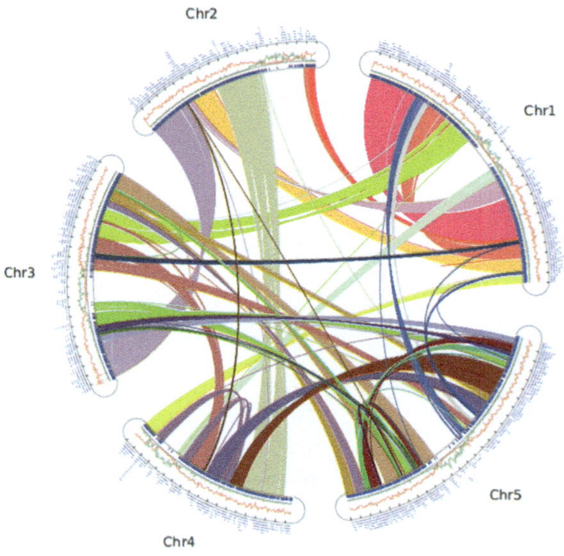

Fig. 3.2 A map anchoring *Brassica oleracea* contigs onto Arabidopsis chromosomes. Chromosomes are arranged in curvy boxes, in which red curves show gene densities, green curves show repetitive sequences density along them, and blue tick marks show distribution of overgo probes. The exterior light-blue blocks show the distribution of syntenic blocks between Arabidopsis chromosomes and Brassica contigs. Lines between chromosomes link the syntenic genes in Arabidopsis genome, and colors distinguish different duplicated blocks

We failed to map 950 *B. oleracea* contigs, accounting for 33% of the total and involving 16% (7,240) BACs, onto Arabidopsis genomic sequences. We hypothesized that most of these failing contigs were pericentromere-related. The hypothesis got support by three different lines of evidence. Firstly, they were found to be gene-scarce: less than 1% (25) gene-derived probes were involved in these contigs. Secondly, 46% of the involved BACs were inferred to be repeat-related, just comparable to those contigs aligned to the Arabidopsis heterochromatin, much higher than those aligned to the euchromatin (34%) (*P*-value = 0). Thirdly, when BES were searched against two Brassica-centromere-specific repeats (CentrBr1 and CentBr2, and each is 176 bp), it was found that these non-anchored contigs were likely centromere-related (18%), very similar to the heterochromatin-related ones (19%), and much more than the euchromatin-related ones (10%).

The regions related to the largest homology depths were not likely RNA-related but possibly related to transposons, while RNA-related contigs were linked mainly to two pericentromeric regions on Arabidopsis chromosomes 2 and 3, showing possible expansion of their related orthologous copies in *Brassica*.

3.3.5 Evolution of Centromeric Repeats

With both *B. oleracea* and *B. rapa,* many CentBr1 and CentBr2 repeats were found in the BESs, permitting comparative analysis to understand their evolution. Actually, 791 and 563 *B. oleracea* BESs are CentBr1- and CentBr2-related, respectively, accounting up to 2% of the total BESs. Comparatively, 20% of *B. rapa* BESs were found to be linked to the above two types of centromere repeats. They showed a15-times difference in the two above repeat types (17,156 vs. 1,132). In both species, around 50% of the BACs were likely from one repeat class in both ends of each BAC, while only a small fraction ($\sim$0.5%) linked to both repeats, showing their separate distribution and expansion in the studied genomes. Although the two repeat classes were named according to *B. rapa*,

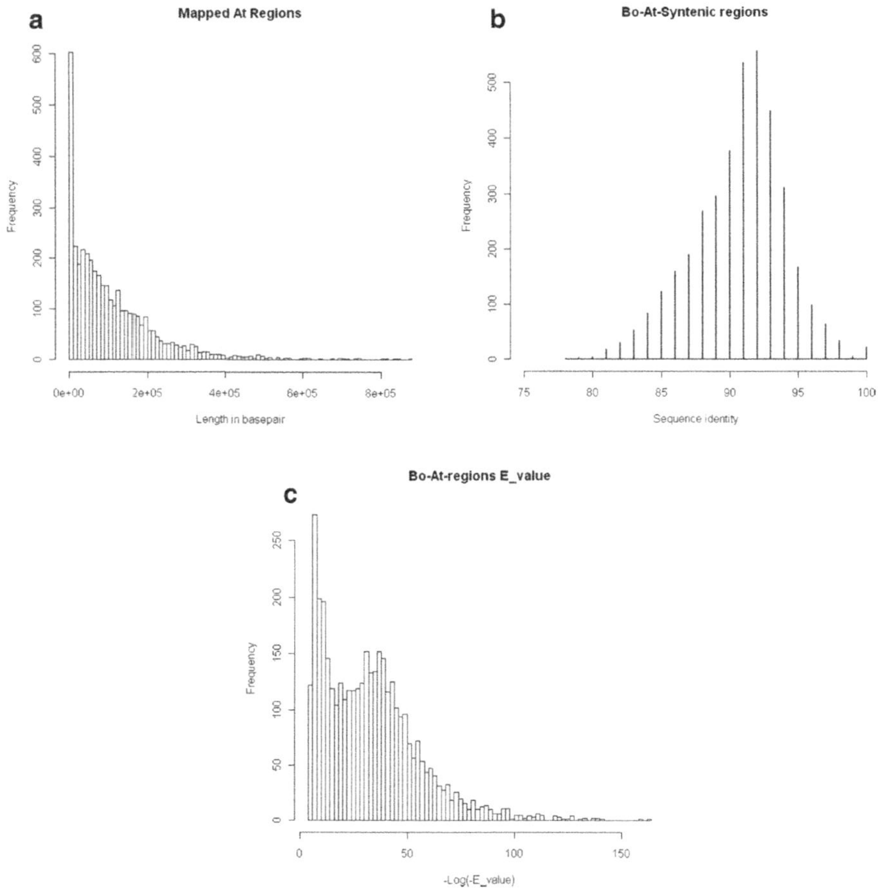

Fig. 3.3 Histograms of mapping *B. oleracea* (Bo) contigs onto *A. thaliana* (At) genome. **a** Size of anchored regions in Arabidopsis genome sequences; Sequence similarity (**b**) and BLAST E-values **c** between anchored Bo and At sequences

the wide-spread distribution of them in both Brassica plants showed that they are likely common to the genus.

We revealed 2,894 and 62,222 of two repeat classes, respectively. By using a random selection process, we obtained 100 *B. oleracea* sequences and 200 *B. rapa* sequences, and performed a phylogenetic analysis (Fig. 3.4). As anticipated, the CentBr2 repeats were grouped together to form a subtree, where the repeats from each species formed two subgroups, each being with repeats from the other species. This showed their separate divergence and expansion in each species. The CentBr1 repeats from both species were mixed with each other, but forming certain subclusters to support their separate expansion. Taken together, our findings suggested the origination and divergence in the Brassica common ancestor.

3.3.6 Mapping onto the Other Eudicots' Genomes

We successfully mapped 324 (11.2%), 337 (11.6%), and 254 (8.7%) contigs onto papaya, poplar, and grape genome sequences. Averagely, the anchored regions extended 82.9, 87.6, and 85 Kb, and cross-species DNA similarities were 87.8%, 87.6%, and 87.3%, respectively. These findings showed very similar divergence levels from Brassica to the other eudicots, in spite of the

Fig. 3.4 Phylogeny of centromeric repeats. CentBr2 repeats are circled out with a blue square. The *B. oleracea* repeats are displayed with red branches and *B. rapa* repeats with black branches

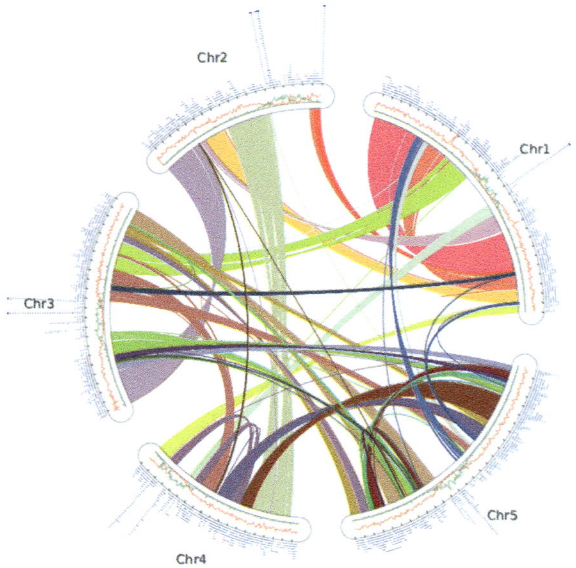

Fig. 3.5 A map of anchoring *Brassica rapa* contigs onto Arabidopsis chromosomes. Chromosomes are arranged in curvy boxes, with red curves to show gene densities, green curves show repetitive sequences density along them, and blue tinkers show distribution of overgo probes. The exterior light-blue blocks show the distribution of syntenic blocks between Arabidopsis chromosomes and *B. rapa* contigs. Blocks accumulating to more than 20 layers are not shown and are all repeat-related ones, the number of which is much larger than in *B. oleracea*, possibly a result of overrepresentation in *B. rapa*, for extra-efforts had been performed to find repeat-related BACs. Lines between chromosomes link the syntenic genes in arabidopsis genome, and colors distinguish different duplicated blocks

divergence times were much divergent, with papaya being in the order Brassicales and the others being much more distant.

3.4 Impacts of Physical Mapping

3.4.1 Synteny Preservation and Recombination

According to our observation above, most Brassica contigs and BACs, involving the majority of probes, were related to the Arabidopsis euchromatin regions. This suggests, after 15–20 Million years of divergent evolution, chromosomal synteny has been much preserved between Brassica and Arabidopsis. In comparison, few contigs could be related to the heterochromatin regions and non-anchored contigs were quite likely related to repetitive sequences, especially centromere-related ones, showing Brassica (peri)centromeric regions were composed of heterochromatic DNA and retain much less synteny to Arabidopsis.

The above findings also suggested well preserved paralogous DNA synteny produced by polyploidization (Fig. 3.2). Besides, the more distant from the pericentromeres of Arabidopsis chromosomes, the more deep is the coverage by Brassica contigs. Lower synteny in (peri)centromeric regions implied that they had been much more fractionated than the euchromatic regions after Brassica-Arabidopsis divergence.

One of the findings from comparison of the sorghum physical map to the rice genome sequence was that regions in which recombination was relatively abundant preserved gene order across taxa. In consideration of the above findings in view of Muller's ratchet (Muller 1964), one would predict most rearrangements to be somewhat deleterious, in that gene arrangement appears to be much more strongly preserved in recombinogenic than in non-recombinogenic regions such as pericentromeric space.

3.4.2 Contribution to the Effort to Sequencing

Recurrent polyploidization makes it difficult to reliably assemble genome sequences, which highlight the importance of constructing an independent physical map for each Brassica species. Based on our findings and those in previous publications, there are many duplicated blocks, making Brassica genomes very complex to decipher. The possibility of homeologous recombination makes things even more complex. Though the frequency of occurrence is very low in one generation (Udall et al. 2005), accumulating effects through many generations may be high. This may hinder efforts to reconstruct the evolutionary history of genes or genome structures. Gene conversion or homeologous DNA translocation could keep two homeologous DNA segments very similar. Together with accumulation repetitive sequences, this will make sequence assembly difficult, especially for those efforts adopting only the second but not the third generation sequencing technologies for budget limit, which produce relatively shorter reads.

References

Babula D, Kaczmarek M, Barakat A, Delseny M, Quiros CF, Sadowski J (2003) Chromosomal mapping of *Brassica oleracea* based on ESTs from Arabidopsis thaliana: complexity of the comparative map. Mol Genet Genom 268:656–665

Blanc G, Hokamp K, Wolfe KH (2003) A recent polyploidy superimposed on older large-scale duplications in the *Arabidopsis* genome. Genome Res 13:137–144

Bowers JE, Arias MA, Asher R, Avise JA, Ball RT et al (2005) Comparative physical mapping links conservation of microsynteny to chromosome structure and recombination in grasses. Proc Natl Acad Sci USA 102:13206–13211

Bowers JE, Chapman BA, Rong J, Paterson AH (2003) Unravelling angiosperm genome evolution by phylogenetic analysis of chromosomal duplication events. Nature 422:433–438

Chalhoub B, Denoeud F, Liu S, Parkin IA, Tang H et al (2014) Plant genetics. Early allopolyploid evolution in the post-Neolithic *Brassica napus* oilseed genome. Science 345:950–953

De Bodt S, Maere S, Van de Peer Y (2005) Genome duplication and the origin of angiosperms. Trends Ecol Evol 20:591–597

Jaillon O, Aury JM, Noel B, Policriti A, Clepet C et al (2007) The grapevine genome sequence suggests ancestral hexaploidization in major angiosperm phyla. Nature 449:463–467

Johnston JS, Pepper AE, Hall AE, Chen ZJ, Hodnett G, Drabek J, Lopez R, Price HJ (2005) Evolution of genome size in Brassicaceae. Amer J Bot 95:229–235

Lan TH, DelMonte TA, Reischmann KP, Hyman J, Kowalski SP, McFerson J, Kresovich S, Paterson AH (2000) An EST-enriched comparative map of Brassica oleracea and *Arabidopsis thaliana*. Genome Res 10:776–788

Lewin HA, Larkin DM, Pontius J, O'Brien SJ (2009) Every genome sequence needs a good map. Genome Res 19:1925–1928

Lim KB, Yang TJ, Hwang YJ, Kim JS, Park JY et al (2007) Characterization of the centromere and pericentromere retrotransposons in *Brassica rapa* and their distribution in related Brassica species. Plant J 49:173–183

Liu S, Liu Y, Yang X, Tong C, Edwards D et al (2014) The *Brassica oleracea* genome reveals the asymmetrical evolution of polyploid genomes. Nat Comm 5:3930

Lukens L, Zou F, Lydiate D, Parkin I, Osborn T (2003) Comparison of a *Brassica oleracea* genetic map with the genome of *Arabidopsis thaliana*. Genetics 164:359–372

Lysak MA, Koch MA, Pecinka A, Schubert I (2005) Chromosome triplication found across the tribe Brassiceae. Genome Res 15:516–525

Martin W (2003) Gene transfer from organelles to the nucleus: frequent and in big chunks. Proc Natl Acad Sci USA 100:8612–8614

Ming R, Hou S, Feng Y, Yu Q, Dionne-Laporte A et al (2008) The draft genome of the transgenic tropical fruit tree papaya (*Carica papaya* Linnaeus). Nature 452:991–996

Muller HJ (1964) The relation of recombination to mutational advance. Mutat Res 1:2–9

Mun JH, Kwon SJ, Yang TJ, Kim HS, Choi BS et al. (2008) The first generation of a BAC based physical map of *Brassica rapa*. BMC genomics 9.

Parkin IA, Sharpe AG, Lydiate DJ (2003) Patterns of genome duplication within the *Brassica napus* genome. Genome 46:291–303

Paterson AH, Bowers JE, Chapman BA (2004) Ancient polyploidization predating divergence of the cereals, and its consequences for comparative genomics. Proc Natl Acad Sci USA 101:9903–9908

Paterson AH, Lan TH, Amasino R, Osborn TC, Quiros C (2001) Brassica genomics: a complement to, and early beneficiary of, the Arabidopsis sequence. Genome Biol 2: REVIEWS1011.

Schmidt R, Acarkan A, Boivin K (2001) Comparative structural genomics in the Brassicaceae family. Plant Physiol Biochem 39:253–262

Schranz ME, Lysak MA, Mitchell-Olds T (2006) The ABC's of comparative genomics in the Brassicaceae: building blocks of crucifer genomes. Trends Plant Sci 11:535–542

Soderlund C, Humphray S, Dunham A, French L (2000) Contigs built with fingerprints, markers, and FPC V4.7. Genome Res 10:1772–1787

Soltis DE, Buggs RJ, Barbazuk WB, Schnable PS, Soltis PS (2009) On the origins of species: does evolution repeat itself in polyploid populations of independent origin? Cold Spring Harb Symp Quant Biol. https://doi.org/10.1101/sqb.2009.74.007[pii]

Tang H, Bowers JE, Wang X, Ming R, Alam M, Paterson AH (2008) Synteny and collinearity in plant genomes. Science 320:486–488

The Arabidopsis Genome Initiative (2000) Analysis of the genome sequence of the flowering plant *Arabidopsis thaliana*. Nature 408:796–815

Thon MR, Martin SL, Goff S, Wing RA, Dean RA (2004) BAC end sequences and a physical map reveal transposable element content and clustering patterns in the genome of Magnaporthe grisea. Fungal Genet Biol 41:657–666

Town CD, Cheung F, Maiti R, Crabtree J, Haas BJ et al (2006) Comparative genomics of *Brassica oleracea* and *Arabidopsis thaliana* reveal gene loss, fragmentation, and dispersal after polyploidy. Plant Cell 18:1348–1359

Tuskan GA, Difazio S, Jansson S, Bohlmann J, Grigoriev I et al (2006) The genome of black cottonwood, *Populus trichocarpa* (Torr. & Gray). Science 313:1596–1604

Udall JA, Quijada PA, Osborn TC (2005) Detection of chromosomal rearrangements derived from homeologous recombination in four mapping populations of *Brassica napus* L. Genetics 169:967–979

Wang X, Shi X, Hao B, Ge S, Luo J (2005) Duplication and DNA segmental loss in the rice genome: implications for diploidization. New Phytol 165:937–946

Wang X, Tang H, Bowers JE, Paterson AH (2009) Comparative inference of illegitimate recombination between rice and sorghum duplicated genes produced by polyploidization. Genome Res 19:1026–1032

Wang X, Wang H, Wang J, Sun R, Wu J et al (2011) The genome of the mesopolyploid crop species *Brassica rapa*. Nat Genet 43:1035–1039

Wendel JF (2000) Genome evolution in polyploids. Plant Mol Biol 42:225–249

Whole-Genome Sequencing, Assembling, and Annotation of *Brassica oleracea* Genomes

4

Ning Guo, Guusje Bonnema, Shenyun Wang, and Fan Liu

Abstract

A high-quality reference genome is critical for understanding genome structure, genetic variation, and evolution of a species. Of the three available *Brassica oleracea* genomes, cabbage 'Cap02-12' and kale-like rapid-cycling 'TO1000' were sequenced and assembled by short reads, hindering assembly through complex repeat regions, while broccoli 'HDEM' was assembled using Nanopore long reads combined with optical mapping. In this chapter we present the de novo genome assembly of cauliflower inbred line 'Korso' using a strategy combining PacBio long reads, optical mapping, and chromosome-contact maps (Hi-C). The assembly and annotation results of cauliflower 'Korso' are compared with the different genome assemblies published before. *B. oleracea* genome assemblies of cauliflower and broccoli, generated by new long-range technologies (PacBio and Nanopore) were longer, completer, and more continuous, with fewer errors, with improved annotation protein-coding genes, transposon distributions, and centromere positions and sequences. Compared to all other three assemblies, more repeat elements were predicted in 'Korso', and the arrangement of the genomic blocks representing the three subgenomes was improved. The availability of new morphotype specific high-quality reference genomes offers an unprecedented opportunity for extensive comparison of intraspecific genome diversity in *B. oleracea*.

N. Guo (✉) · F. Liu (✉)
Key Laboratory of Biology and Genetic Improvement of Horticultural Crops (North China), Beijing Vegetable Research Center, Beijing Academy of Agriculture and Forestry Sciences, National Engineering Research Center for Vegetables, The Ministry of Agriculture and Rural Affairs, Beijing 100097, China
e-mail: guoning@nercv.org

F. Liu
e-mail: liufan@nercv.org

G. Bonnema
Plant Breeding, Wageningen University and Research, 6708 PB Wageningen, The Netherlands
e-mail: guusje.bonnema@wur.nl

S. Wang
Vegetable Research Institute, Jiangsu Academy of Agricultural Science, Nanjing 210014, China
e-mail: wangshenyun@jaas.ac.cn

4.1 Introduction

4.1.1 The Development of Sequencing Technologies

Forty years ago, the complete genome sequences can be deciphered for the first time due to the advent of Sanger sequencing. The next-generation sequencing (NGS) technologies achieved the

S. Liu et al. (eds.), *The Brassica oleracea Genome*, Compendium of Plant Genomes,
https://doi.org/10.1007/978-3-030-31005-9_4

second revolution to made genome sequencing much cheaper and faster. The advent of NGS technologies reduced the costs of sequencing substantially, which has enabled sequencing of over 100 plant genomes. However, NGS methods have several drawbacks and disadvantages, most notably their short reads. The assemblies based on short reads generated by NGS platforms are highly fragmented with tens of thousands of short contigs distributed in thousands of scaffolds, and are inaccurate with inherent biases and nonrandom sequencing errors (Ross et al. 2013; Michael and VanBuren 2015). These genome assemblies are missing biologically meaningful sequences including entire genes, regulatory regions, transposable elements, centromeres, telomeres, and haplotype-specific structural variations, complicating downstream applications for comparative genomics, gene discovery, genome structural genetic variation, and evolutionary analysis. In addition, it is hard for the short-read technologies to assembling repeat elements, which blocks the studies to understand the dynamics of transposable elements. Recently, third-generation (long-read) sequencing technologies appeared which can produce genome assemblies of unprecedented quality. Moreover, these technologies can directly detect epigenetic modifications of native DNA and allow whole-transcript sequencing without the need for assembly. This marks the third revolution in sequencing technology (Van Dijk et al. 2018).

4.1.2 Advantages of Third-Generation Sequencing Technologies

The development of long-read sequencing technologies, such as Pacific Biosciences (PacBio) and Oxford Nanopore Technology (ONT) has been used to change the above situation (VanBuren et al. 2015; Chin et al. 2016; Jiao and Schneeberger 2017; Michael et al. 2018). Single-molecule real-time (SMRT) sequencing (Pacific Biosciences) produces long and unbiased sequences, which enables the assembly of complex repeat structures and GC- and AT-rich regions that are often unassembled or highly fragmented in NGS-based draft genomes (Alice 2010; Seung Chul et al. 2013). Scaffolding technologies have been developed for ordering and orienting assembly contigs, including optical mapping (Staňková et al. 2016) and chromosome interaction mapping (Hi-C) (Burton et al. 2013), which provide relatively inexpensive and high-resolution scaffolding and chromosome anchoring results. Optical mapping technologies observe the linear separation of small DNA motifs (often restriction enzyme recognition sites or nickase sites), which can provide sufficient contextual information to scaffold assembled contigs or correct existing reference assemblies. Hi-C is an adaptation of the chromosome conformation capture (3C) methodology that identifies long-range chromosome interactions in an unbiased fashion without a priori target site selection. The frequency of long-range consensus interactions decays rapidly as a linear distance along with a chromosome increases, allowing Hi-C data to scaffold assembled contigs to the scale of full chromosomes. Several high-quality genome assemblies, some were firstly sequenced and some were upgrades, have been published using single-molecule sequencing independently, or combined with optical mapping and/Hi-C technologies (Du et al. 2017; Jiao et al. 2017; Michael et al. 2018; Sun et al. 2018).

4.1.3 The *Brassica* Genus and *B. oleracea* Species

The genus *Brassica* comprises various economically important species, many of which are extensively cultivated as oil crops and vegetables all over the world. *Brassica oleracea* is one of three diploid *Brassica* species in the classical triangle of U (Nagaharu 1935) that also includes diploids *B. rapa* (AA) and *B. nigra* (BB) and allotetraploids *B. juncea* (AABB), *B. napus* (AACC) and *B. carinata* (BBCC). *B. oleracea* comprises many important worldwide vegetable crops including cauliflower, broccoli, cabbage, Brussels sprout, kohlrabi, and kale, demonstrating extreme morphological diversity in their harvested organs, like inflorescences, leaves, axillary buds, and stems. Cauliflowers are characterized by their curds, which are composed of primary inflorescence meristems with floral primordia arrested in their development, and cabbages by their leafy heads. *B. oleracea* vegetables are excellent sources of natural antioxidants and anti-carcinogens in diets due to their high levels of various active phytochemicals such as diverse phenolic compounds and glucosinolates (GSLs) (Halkier and Gershenzon 2006). *B. oleracea* has undergone a Brassicaceae-lineage-specific whole-genome triplication (WGT) since its divergence from the Arabidopsis lineage (Bowers et al. 2003; Lysak et al. 2005). This was followed by diploidization with substantial genome reshuffling and gene loss (Wang et al. 2011), which makes *B. oleracea* a model for the study of polyploid genome evolution, mechanism of duplicated gene loss, neo- and sub-functionalization and associated impact on morphological diversity and species differentiation (Cheng et al. 2016).

4.1.4 The Sequenced *B. oleracea* Genomes

Complete and accurate reference genomes and annotations provide fundamental tools for the characterization of genetic and functional variations. Most of the published reference genomes of *Brassica* species are sequenced using short-read strategies, including two genomes of *B. oleracea*: one is heading cabbage 'Cap02-12' (Liu et al. 2014), another is a kale-like rapid-cycling type 'TO1000' (Parkin et al. 2014). These genomes are highly fragmented, with many gaps, ambiguities, and errors. About 20–30% of genome sequences are not assembled and the ratio anchored to chromosomes is even lower. Although these initial references enabled the rapid progress in *Brassica* genomics, the original assembly is composed of more than tens of thousands small contigs, many of which are arbitrarily ordered and oriented, markedly complicating detailed analysis of individual loci and impeding the investigation of intergenic regions crucial to our understanding of phenotypic variation and genome evolution. With these genomes assembled using NGS, it is very hard to identify the genomic structural variants and to elucidate the genomic bases of morphological variations and domestication. Recently, an upgraded Chinese cabbage (*B. rapa*) genome has been sequenced and assembled using PacBio combined with Hi-C (Lei et al. 2018), while a sarson type (*B. rapa*) and broccoli (*B. oleracea*) genome have been generated using nanopore long reads and optical maps (Belser et al. 2018).

4.2 Genome Assembly of *B. oleracea*

4.2.1 Assembly Strategies of Published *B. oleracea* References

As mentioned above, there are three available *Brassica oleracea* genomes: 'Cap02-12' (cabbage, *B. oleracea* ssp. *capitata*) (Liu et al. 2014), 'TO1000' (*B. oleracea* kale-like type) (Parkin et al. 2014), and 'HDEM' (Broccoli, *B. oleracea* ssp. *botrytis italica*) (Belser et al. 2018). 'Cap02-12' and 'TO1000' were sequenced using short-read strategies, but the resulting assemblies are fragmented (contig N50 of <50 kb) and contain a high proportion of unknown or missing bases (22–30%). For 'HDEM', a sequencing strategy was used combining MinION (Oxford Nanopore), optical maps by the Saphyr system (BioNano Genomics), and short reads from an Illumina sequencer.

The draft genome assembly of *B. oleracea* var. *capitata* line 'Cap02-12' was produced by Illumina, Roche 454, and Sanger sequence data. This assembly represented 85% of the estimated 630 Mb genome with contig and scaffold N50 26.8 Kb and 1.46 Mb respectively, including >98% of the gene space. The assembly was anchored to a genetic map to produce nine pseudo-chromosomes that account for 72% of the assembly (Liu et al. 2014).

The genome assembly of the double haploid *B. oleracea* kale-like type 'TO1000DH' was generated from approximately 94X coverage of the estimated 648 Mb genome. A hybrid assembly approach combining Sanger sequencing reads, Illumia and Roche 454 sequencing data was taken. The final assembly totaling 488.6 Mb covered approximately 75% of the estimated genome size (648 Mb) and was arranged into 33,459 scaffolds (2172 of which were greater than 2 Kb), with an N50 of 850 Kb. Almost 94% of the assembly could be assigned to one of the nine pseudo-chromosomes of *B. oleracea* by anchoring of the scaffolds to the linkage map (Parkin et al. 2014).

The genome of broccoli ('HDEM' genotype, estimated size of 630 Mb) was sequenced by a strategy combining MinION, which is a portable sequencer commercialized by the Oxford Nanopore company, optical maps produced using the Saphyr system (Bionano Genomics), and short reads from an Illumina sequencer. The nanopore long reads and long-range information provided by the optical maps were used to get the final assembly with N50 9.5 and 29.5 Mb at the contig and scaffold levels, respectively. About 95.3% of the assembly was anchored to the nine chromosomes (Belser et al. 2018). Compared with the two assemblies ('Cap02-12' and 'TO1000') generated by NGS short reads, the contig N50 of 'HDEM'assembly by Nanopore long reads combined with optical mapping data were ~400-times higher.

4.2.2 The Sequencing and Assembly of a Cauliflower Reference Genome

Recently, we de novo sequenced the genome of a cauliflower (*B. oleracea* ssp. *botrytis* 'Korso') with a sequencing strategy combining single-molecule real-time sequencing (SMRT, PacBio), optical maps produced using the Saphyr system (BioNano Genomics), chromosome-contact maps (Hi-C), and short reads from an Illumina sequencer (unpublished). Using high-quality long fragment DNA from the leaves of 'Korso', we generated ~120-fold coverage of long-read sequencing data (70.0 Gb; 8801,757 subreads with mean and N50 length of 8.0 Kb and 13.1 Kb, respectively) using 24 single-molecule real-time (SMRT) cells on the PacBio RSII and Sequel platform. Initially, we assembled the long reads from PacBio SMRT sequencing alone by using the PacBio diploid assembler software FALCON (https://github.com/PacificBiosciences/FALCON). FALCON started with the step of filtering SMRT reads with parameter (Options–filter = 'MinReadScore = 0.80, MinSRL = 500, MinRL = 100'), and then performed error correction and pre-assembly and contig construction with parameters (the length_cutoff and length_cutoff_pr set to 13 kb and 14 kb, respectively). The assembled contigs were polished with Pilon using 100.12Gb high-quality Illumina paired-end reads. In addition, a genome map was assembled from 242.2 Gb cleaned BioNano optical map data and used to connect the assembled contigs. We generated a genome assembly of 'Korso' comprising 615 contigs, with cumulative lengths of 549.7 Mb and N50 sizes of 4.97 Mb (Table 4.1). Furthermore, 285.8 million cleaned Hi-C read pairs, among which 58.6 million were valid, were used for pseudo-chromosome construction by HiC-Pro and LACHESIS software. Finally, 477 high-confident contigs were anchored to a total of nine chromosomes accounting for 547.56 Mb (99.04% of all the assembly) and leaving 138 very short contigs unanchored totaling 5,291,193 bp (Tables 4.1 and 4.2). The assembly process for producing chromosome-length scaffolds is shown in Fig. 4.1. The final assembly totaling 552.8 Mb covered approximately 97.5% of the estimated genome size of 566.9 Mb, which was estimated by flow cytometry using *B. rapa* as reference.

4.3 Whole-Genome Comparison of *B. oleracea* Reference Genomes

Comparing the assemblies of *B. oleracea* genomes (cabbage 'Cap02-12', kale-like rapid-cycling 'TO1000', broccoli 'HDEM', and cauliflower 'Korso'), the two reference genomes sequenced by new third-generation sequencing

Table 4.1 Statistics of the genome assemblies of *B. oleracea*

	Cap02-12	TO1000	HDEM	Korso
References	Liu et al.	Parkin et al.	Belser et al.	Unpublished
Estimated genome size (Mb)	630	648	630	567
Number of scaffold (≥ 2 Kb)	1809	1428	140	10
Cumulative size	539,907,250	473,834,292	554,975,960	552,847,295
N50 (L50)	1,457,055 (224)	48,366,697 (5)	29,516,207 (8)	59,208,503 (5)
N90 (L90)	258,906 (388)	39,822,476 (9)	13,883,733 (17)	49,705,675 (9)
Maximum size	8,788,225	64,984,695	48,260,371	73,605,601
Number of contigs (≥ 500 bp)	167,493	51,566	264	615
Cumulative size	502,114,421	435,858,618	545,017,856	549,672,019
N50	26,828 (5.425)	22,128 (5,797)	9,491,203 (19)	4,972,625 (26)
N90	3,527 (22,669)	4,448 (21,523)	2,202,317 (59)	1,717,536 (90)
Maximum size	199,461	163,976	26,712,175	28,617,232

Table 4.2 The length of pseudo-chromosomes of different assemblies of *B. oleracea*

Pseudo-chromosomes	Length			
	Korso	HDEM	TO1000	Cap02-12
C01	53,272,422	52,030,199	43,754,388	38,761,720
C02	63,615,344	63,782,621	52,875,895	44,046,003
C03	78,655,145	73,711,317	64,974,595	57,781,463
C04	68,427,143	65,381,821	53,705,393	40,895,475
C05	57,510,194	54,996,917	46,892,785	32,828,328
C06	49,705,675	46,538,988	39,814,676	40,704,471
C07	59,208,503	58,257,932	48,360,397	48,346,208
C08	52,701,841	50,921,300	41,752,485	41,516,064
C09	64,459,193	63,239,600	54,667,868	40,126,856
Anchored	547,555,460	528,860,695	446,798,482	385,006,588
Unanchored	5,291,193	26,116,365	41,776,129	154,963,662
Total	552,847,295	554,977,060	488,574,611	539,970,250
Anchored percentage (%)	99.04	95.29	91.45	71.30

technologies (Oxford Nanopore and PacBio SMRT) had longer and more complete and continuous assemblies. The contig N50s of 'HDEM' and 'Korso' were 9.5 and 8.2 Mb, which were approximately 400 and 300 times higher than the assemblies by short reads. 'Korso' has the longest assemblies anchored to chromosomes compared to the other three ones, being 18.7 Mb, 100.8 Mb, and 162.6 Mb longer than 'HDEM', 'TO1000', and 'Cap02-12', respectively (Table 4.2).

We analyzed synteny between the assembled pseudo-chromosomes of cauliflower ('Korso') with the broccoli ('HDEM'), rapid-cycling ('TO1000'), and cabbage ('Cap 02-12') genomes by MUMmer software (Delcher et al. 2003). 'Korso' had a good syntenic relationship with 'HDEM', whereas the syntenic relationship with

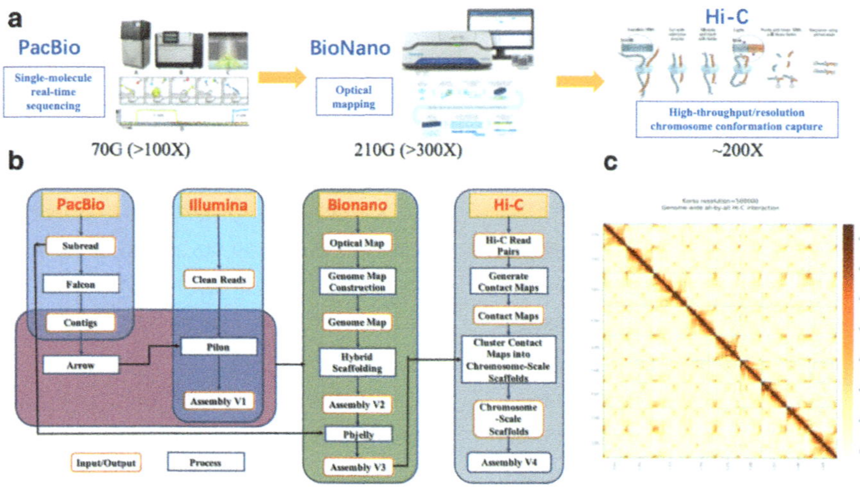

Fig. 4.1 Assembly scheme for producing chromosome-length scaffolds. **a** The flowchart of sequencing strategy for de novo sequencing of cauliflower 'Korso'. **b** The selected assembly method for 'Korso' genome assembly. **c** The Hi-C heatmap for assembly chromosome-length scaffolds

'Cap02-12' was the worst demonstrated by fragmented and translocated colinear blocks (Fig. 4.2). Most of these differences are likely caused by the lower quality of genome assembly using short reads and the anchoring of scaffolds to chromosomes by genetic mapping with many scaffolds containing one or few genetic markers only. 'TO1000' showed better collinearity with 'Korso' than 'Cap 02-12', with still some large common variants between 'Korso' and 'TO1000', and between 'Korso' and 'Cap02-12', such as inversions on chromosome C01, C06, and C09, and deletions on C03, C04, C05, and C08 (Fig. 4.3). When comparing 'Korso' with 'TO1000' and 'Cap02-12' using the aligned regions indicating syntenic blocks bigger than 10 Kb by MUMmer analysis, many structural variations such as insertions/deletions, translocations, and inversions were revealed (Fig. 4.3). Further analysis shows that a part of the variations were deletions in 'TO1000' and 'Cap02-12' which may point to unassembled sequences or unanchored scaffolds, especially at or near the centromeres, such as the deletions on C04 (Fig. 4.3), and the inversions around the centromere region of C01 and C06. Inaccurate linkage relationships around the centromeres of the genetic maps likely caused these discrepancies. Taking the chromosome C01 as an illustration, the centromere regions of the assemblies of 'TO1000' and 'Cap02-12' in this chromosome were almost lost compared to the 'Korso' centromere (Fig. 4.4b, d and e). Near the C01 centromere region, 'TO1000' and 'Cap02-12' both had an inversion in comparison with 'Korso' (Fig. 4.4d and e). In conclusion, the incorrect orientation of scaffolds caused an apparent inversion of 'TO1000' and 'Cap02-12'. These findings, which were verified by the short reads mapping results of 'Korso' (Fig. 4.4a), demonstrated that the assemblies by short reads were incomplete and had a high error rate especially in the highly repetitive sequence region such as centromeres.

4.4 Genomic Annotation of the *B. oleracea* Genome

4.4.1 Gene Annotation for 'Korso'

To annotate the protein-coding genes in the cauliflower 'Korso' genome, we used a method combining RNA-seq-based transcriptomes (from seven different tissues such as root, stem, leaf,

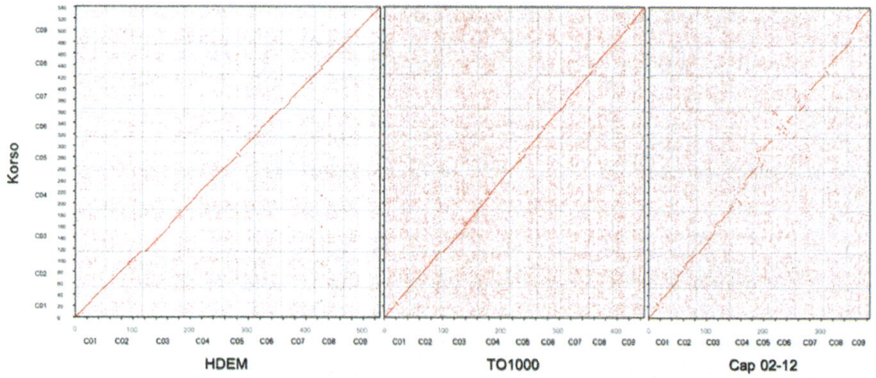

Fig. 4.2 The whole-genome syntenic relationship by MUMmer analysis between 'Korso' and 'HDEM', 'TO1000', 'Cap02-12', respectively

curd, bud, flower, and silique), full-length transcriptomes sequencing (mixed sample of the seven tissues) by PacBio Sequel platform, ab initio gene model prediction, and protein-homology-based prediction to get the complete and accurate gene set of 'Korso', as well as to identify comprehensive transcriptomes by alternative splicing. In total, 60,640 high-confidence protein-coding genes were predicted in the 'Korso' genome, and 70.86% (42,969) of the predicted genes were supported by RNA-seq data from seven different tissues of 'Korso'. In total, 60,027 (98.99%) of the 'Korso' predicted genes were allocated to nine pseudo-chromosomes. The number of annotated genes of 'Korso', 'HDEM', and 'TO1000' were around 60,000, while 'Cap02-12' was annotated to have 45,758 genes. Comparing the annotation results, the average gene length of 'Korso' especially the intron length was longer than that of 'HDEM', 'TO1000', and 'Cap 02-12' (Table 4.3). The gene completeness was improved by long-read assemblies.

Functional assignments of 'Korso' predicted genes were made with BLAST by aligning their protein-coding regions to sequences in the public protein database, including KEGG, COG, GO, Siwssprot, and TrEMBL. The gene numbers annotated in the different database are listed in Table 4.4 and a total of 55,534 genes were functionally annotated, comprising 91.00% of 60,640 gene models (Table 4.4).

4.4.2 Repeat Element Annotation for 'Korso'

Repeat families found in the genome assemblies of 'Korso' were first independently identified de novo and classified using the software RepeatModeler (https://github.com/rmhubley/RepeatModeler) (version 1.0.11). Meanwhile, LTR (long terminal ret-rotransposon) finder and LTRharvest were employed to identify the complete LTR-retrotransposons, and LTR retriver integrated both outputs to produce a nun-redundant completer LTR-retrotransposons set. Finally, the results from Repeat modeler, Ltr retriver, and the sequences from Repbase were merged together as the input of RepeatMasker (http://www.repeatmasker.org/) (version 4.0.7) to discover and identify repeats for 'Korso' genomes.

The 'Korso' genome contains a total of 55.35% repetitive elements. The results of the repeat classification are summarized in Table 4.5. Comparing with previous references, more repeat elements were predicted in 'Korso'. In our novel assembly of 'Korso', the most abundant TEs were LTR-RT, which covered a total length of 172.29 Mb and represented 31.16% of the assembled genome (Table 4.5). Non-LTR-RT repeats accounted for 3.86% of the 'Korso' assembly. The detected DNA transposons covered a total of 61.32 Mb, which made up 11.09% of the genome assembly. The length of Copia and Gypsy LTR-RT were 53.77 and

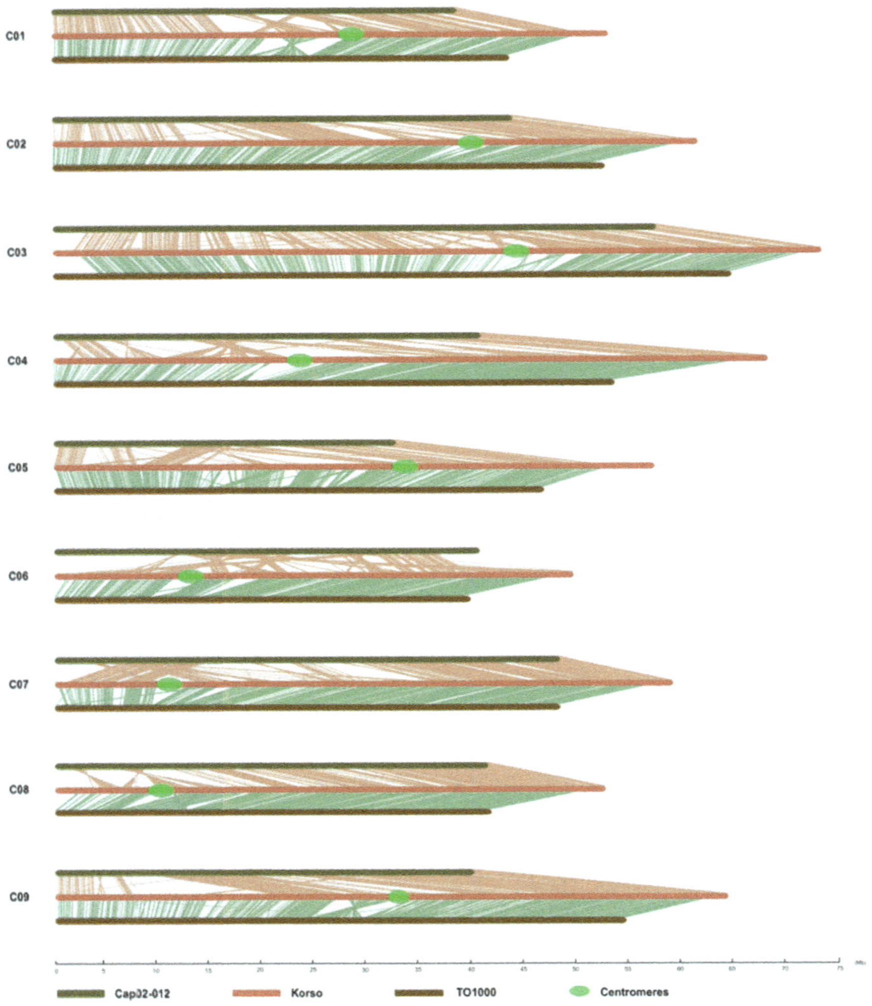

Fig. 4.3 Whole-genome comparison of 'Korso', 'TO1000', and 'Cap02-12'. The aligned regions larger than 10 Kb were represented by the crossing lines between each pair of pseudo-chromosomes. The green ovals indicated the centromeres

77.85 Mb representing 9.72 and 14.08% of the assembly of 'Korso', while Copia and Gypsy represented 12 and 8% ('HDEM'), 10 and 7% ('TO1000'), 12 and 8% ('Cap02-12') in the other *B.oleracea*assemblies. We annotated more Gypsy type LTR-RTs than Copia in the assembly in 'Korso', which was consistent with LTR annotation results of the new *B. rapa* V3.0, whereas the other assemblies of *B. oleracea* annotated more Copia than Gypsy. The total annotated transposable elements in the assemblies of 'Korso', 'HDEM', 'TO1000', and 'Cap02-12' were 45%, 37%, 27%, and 36%, respectively. The long-read assemblies improved the completeness of the transposable element catalogue as well as the genomic context of these transposable element-rich regions.

The distribution of annotated coding genes, repeat elements, Copia and Gypsy types LTR-RTs on nine chromosomes of 'Korso' is demonstrated in Fig. 4.5. We can clearly see that the distributions of coding genes and repeat elements are the opposite. The regions identified as centromeres had a high density of repeat

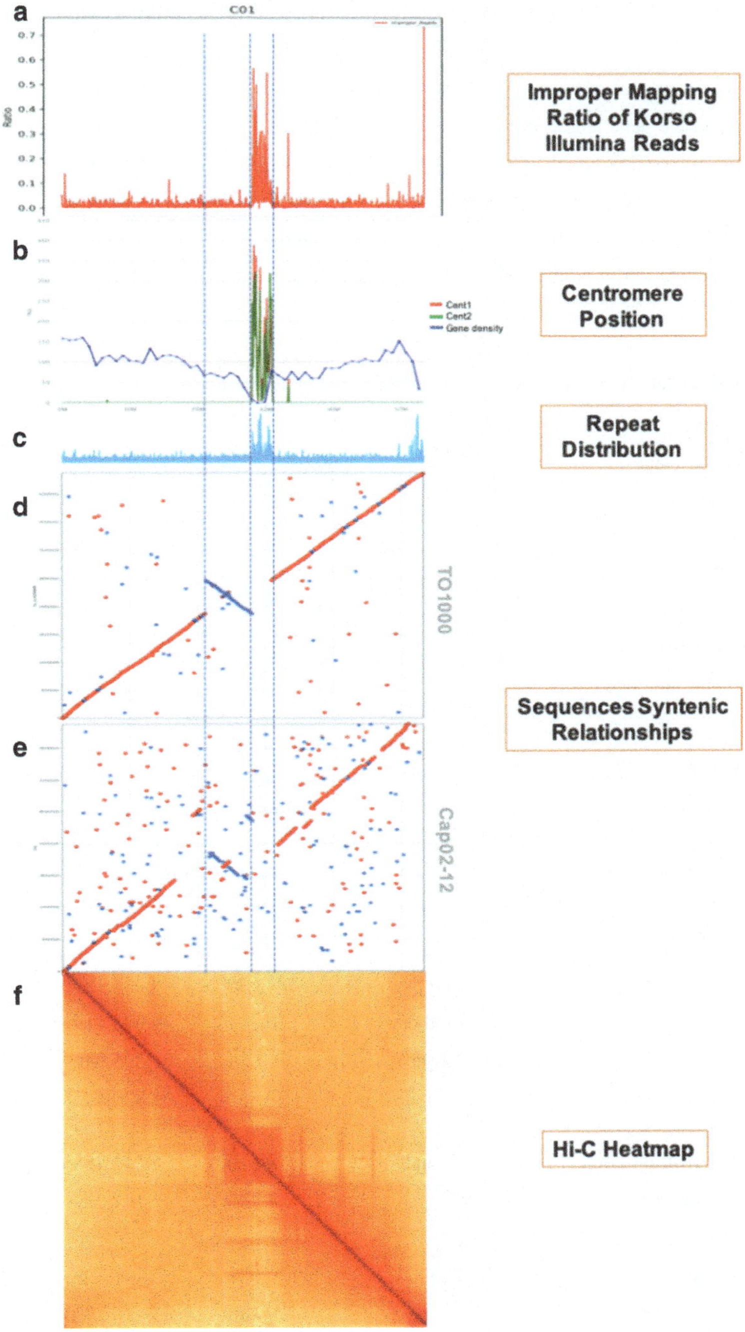

Fig. 4.4 The analysis of the deletions and inversions on chromosome C01 of 'TO1000' and 'Cap02-12'. **a** The improper mapping ratio distribution on 'Korso' chromosome C01 of 'Korso' illumina short reads. **b** The centromere position location of 'Korso' C01 by the distribution of centromere-specific repeats CentBr, and the gene density distribution on C01 of 'Korso'. **c** The repeat sequences distribution on C01 of 'Korso'. **a**, **b**, and **c** demonstrated the location of centromere, and the illumina reads mapping status showed the proper mapping of short reads on C01 except the centromere part because of highly repetitive, which demonstrated the accuracy of assembly. **d** The syntenic relationship between 'Korso' and 'TO1000'. **e** The syntenic relationship between 'Korso' and 'Cap02-12'. **f** The Hi-C heatmap of 'Korso' C01 showed the good assembly

elements and low density of coding genes. The distribution patterns of Copia and Gypsy also differed. We can conclude that due to the high-quality genome assembly and annotation of 'Korso', the distribution of repeat elements over the chromosomes, especially the LTRs distribution on centromeres was improved (Fig. 4.5).

Table 4.3 Statistics of gene annotation of Korso, HDEM, TO1000, and Cap02-12

	Cap02-12	TO1000	HDEM	Korso
Total number of genes	45,758	59,225	61,279	60,640
Average gene length (bp)	1760.73	1749.70	1969.14	2303.61
Average CDS length (bp)	1036.99	1046.79	937.60	931.87
Average exon number per gene	4.55	4.54	4.30	5.74
Average exon length (bp)	228.09	229.80	218.05	213.98
Average intron length (bp)	204.07	200.1	209.46	340.41

Table 4.4 Functional annotation of predicted genes in Korso

Type	Number	Percent (%)
Total Gene	60,640	100
KEGG	19.798	32.00
COG	25,864	42.65
GO	25,148	41.47
Siwssprot	34,783	57.00
TrEMBL	55,414	91.00
Total annotated	55,534	91.00

Table 4.5 Repeat classification in Korso

Repeat class description			Korso				
Repeat types	Order	Superfamily	Total number	Total length (bp)	Intact number	Intact length (bp)	Percent (%)
Class I: retrotransposon	LTR	Total	192843	180840352	161655	172293907	31.16
		Copia	52,996	53,925,706	49,263	53,766,975	9.72
		Gypsy	82,630	79,465,189	71,324	77,849,052	14.08
		Others	76,291	49,993,112	69,947	48,834,748	8.82
	LINE	Total	29,084	21,395,940	28,927	21,386,288	3.86
Class II: DNA transposon	DNA	Total	97,254	61,829,445	93,818	61,318,591	11.09
RC (rolling-circle)		Total	8,510	4,532,484	8509	4,532,469	0.81
Simple_repeat			4,436	5,454,069	4,436	5,454,069	0.98
Satellite			2,024	860,190	2,024	860,190	0.15
Unknown[a]			129,081	52,497,825	129,081	52,497,825	9.49
Total content			519,488	333,426,292	353,206	306,055,073	55.35

Those repeats which take <0.01% in all the three species are not listed in the table
[a]Unknown: repeats that can not be attributed to any known class

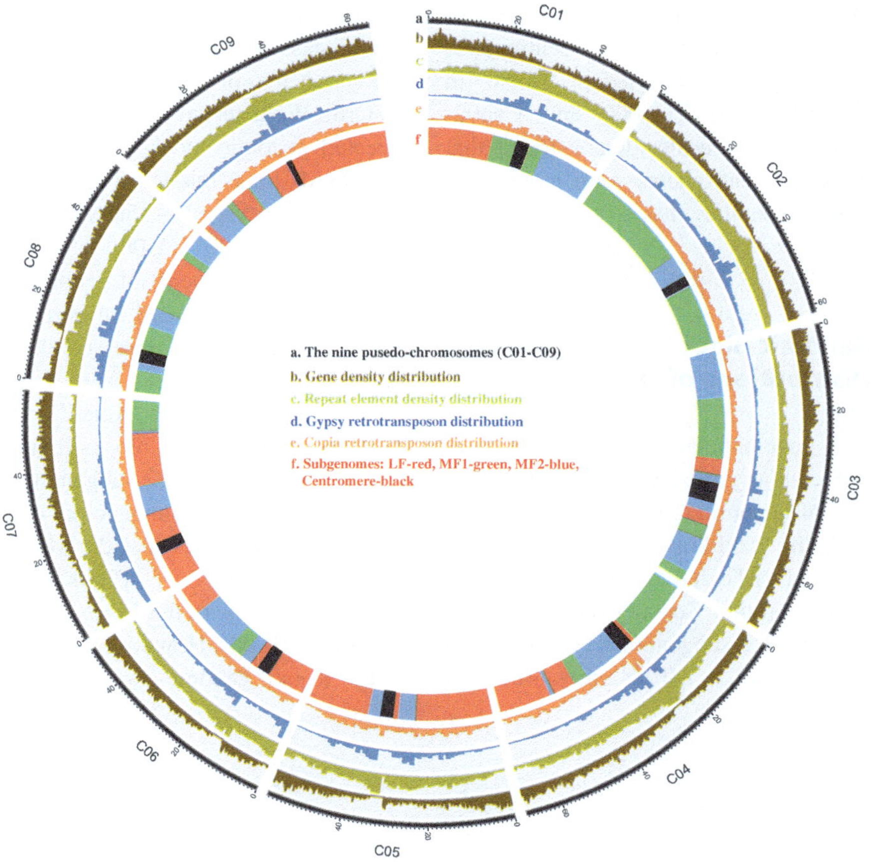

Fig. 4.5 Circos figure showing the distribution of gene density, repeat elements density, Copia and Gypsy type LTRs density, and location subgenomes of *B. oleracea* ssp. *botrytis* 'Korso'

4.5 Centromeres and Genomic Block Distribution in the 'Korso' Genome

4.5.1 Centromere Determination in the 'Korso' Genome

The centromere is responsible for accurate chromosome segregation. Highly repetitive sequences at these centromeres greatly increase the difficulty of the centromere assemblies. Based on the high-quality assembly, we accurately determined the location of the centromeres of all chromosomes in 'Korso' by screening previously determined centromeric repeat sequences of *Brassica*, the centromeric satellite repeats *CentBr1* and *2* (Fig. 4.6). The centromere regions had the lowest gene densities in all the nine chromosomes (Fig. 4.5). To assess the assembly of 'Korso' with regard to the centromeres, we analyzed the sequence features of the centromere regions. The distribution patterns of transposable elements (TEs) support the positions of the centromeres in the genome of 'Korso'. The centromere positions of 'Korso' corresponded very well with the previous cytological centromeres determination results by karyotype analysis using fluorescent in situ hybridization (FISH) (Xiong and Pires 2011). It is well known that TEs are enriched in the flanking regions of centromeres. TE sequences especially Copia and Gypsy type LTRs indeed were present at a relatively high density surrounding the positions of the centromeres of 'Korso' (Fig. 4.5). In addition, more centromere-specific sequences were detected in 'Korso' in comparison to those in 'TO1000' and 'Cap02-12' (Fig. 4.4). These results demonstrated that the assemblies of centromeres were more complete in the assemblies based on long-reads sequencing.

4.5.2 Genomic Block and Subgenome Distributions of 'Korso'

Based on the gene annotation and syntenic relationships of coding genes, we investigated the relationships of genomic blocks using the recent assembly of the cauliflower 'Korso' genome (Cheng et al. 2014). To define the genome blocks and centromeres in the 'Korso' genome, we first constructed the three subgenomes (LF: the least fractionated blocks, MF1: the medium fractionated blocks and MF2: the most fractionated blocks) based on the syntenic relationship between genes of 'Korso' and *A. thaliana* by SynOrth. We detected 71 out of the 72 (3 × 24) expected genomic blocks in 'Korso', as the G block in MF1 was missing (Fig. 4.7). Most of the blocks were arranged in line with the order previously reported in 'TO1000'.

In 'Korso', fragments of genome block D on subgenome LF, identified on chromosome C01 and C04, were not observed in 'TO1000'. Additionally, some blocks were detected in different subgenomes of 'Korso' compared to 'TO1000'. For example, blocks R and W on C02 were present on the MF2 subgenome of 'Korso' while on the MF1 subgenome of 'TO1000'. And R and W blocks in C03 were in MF1 of 'Korso' but in MF2 of 'TO1000'. Also, the J block on C04 was in the LF of 'Korso' while in MF1 of 'TO1000'. However, the block positions of the 'Korso' genome corresponded with those

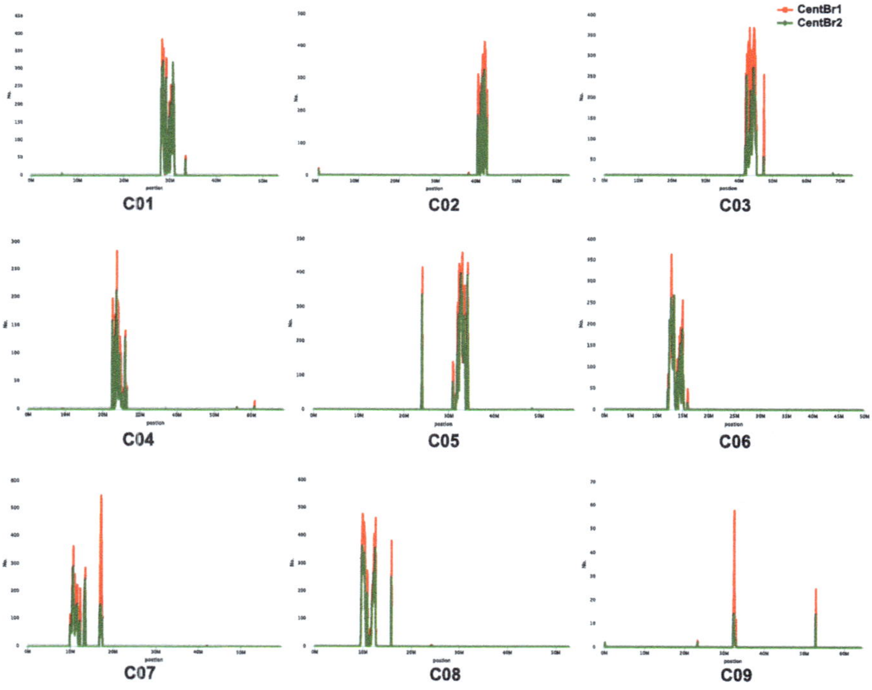

Fig. 4.6 The distributions of centromeric satellite repeats CentBr1 and 2 on the nine chromosomes (C01–C09) of Koros demonstrated the centromere positions. The x axis indicates the position of centromeres of Korso. And the y axis showed copy numbers of CentBr1 and 2. The CentBr1 and 2 were showed in red and green lines, respectively

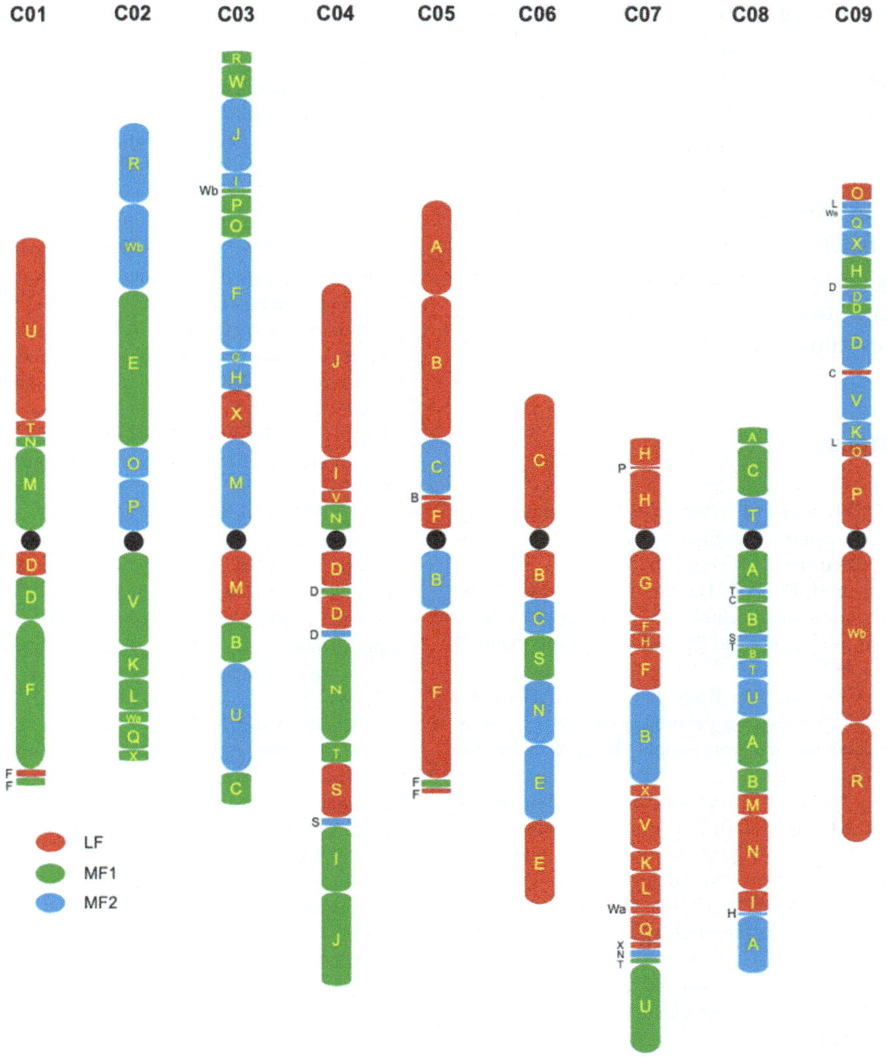

Fig. 4.7 Derived ancestral block structure for Korso. The red, green and blue long ovals represent the genomic blocks on LF, MF1 and MF2 subgenomes respectively. The black circles indicate the centromeres

in the recently assembled *B. rapa* V3.0, which indicates that the block structure of 'Korso' is more complete and accurate than that of 'TO1000'.

4.6 Conclusions

In summary, we compared the recently assembled broccoli genome 'HDEM' (Nanopore and Bionano) and our new assembly of cauliflower 'Korso' (PacBio, Bionao and Hi-C) with 'TO1000' and 'Cap02-12', two references assembled by short reads. The assemblies of 'Korso' and 'HDEM' were longer, more continuous, complete, and accurate, and showed superior quality in repetitive regions, especially the centromere regions. Based on the high-quality assembly, the genome of 'Korso' was annotated using a method combining RNA-seq, full-length transcriptomes, ab initio, and protein homology prediction. The completeness of predicted coding genes of 'Korso' was improved compared to the other *B. oleracea* genomes. More repeat elements were annotated and their distributions were displayed. The integrity of the centromeres was much higher in the assembly of 'Korso'. The improved assembly also made it possible to generate an updated structure of genomic blocks and subgenomes. The improved assembly provides extra resources for gene discovery and for studying genetic variations and genome evolution in *B.oleracea*.

References

Alice MC (2010) Third generation DNA sequencing: pacific biosciences' single molecule real time technology. Chem Biol 17(7):675–676

Belser C, Istace B, Denis E, Dubarry M, Aury JM (2018) Chromosome-scale assemblies of plant genomes using nanopore long reads and optical maps. Nat Plants 4 (11):879–887

Bowers JE, Chapman BA, Rong J, Paterson AH (2003) Unravelling angiosperm genome evolution by phylogenetic analysis of chromosomal duplication events. Nature 422(6930):433

Burton JN, Andrew A, Patwardhan RP, Ruolan Q, Kitzman JO, Jay S (2013) Chromosome-scale scaffolding of de novo genome assemblies based on chromatin interactions. Nat Biotechnol 31(12):1119

Cheng F, Wu J, Wang X (2014) Genome triplication drove the diversification of *Brassica* plants. Hort Res 1(1):14024

Cheng F, Sun R, Hou X, Zheng H, Zhang F, Zhang Y et al (2016) Subgenome parallel selection is associated with morphotype diversification and convergent crop domestication in *Brassica rapa* and *Brassica oleracea*. Nat Genet 48(10):1218

Chin C-S, Peluso P, Sedlazeck FJ, Nattestad M, Concepcion GT, Clum A et al (2016) Phased diploid genome assembly with single-molecule real-time sequencing. Nat Methods 13(12):1050

Delcher AL, Salzberg SL, Phillippy AM (2003) Using MUMmer to identify similar regions in large sequence sets. Curr Prot in Bioinfor 00(1):10.3.1–10.3.18

Du H, Yu Y, Ma Y, Gao Q, Cao Y, Chen Z et al (2017) Sequencing and de novo assembly of a near complete indica rice genome. Nat Commun 8:15324

Halkier BA, Gershenzon J (2006) Biology and biochemistry of glucosinolates. Ann Rev Plant Biol 57:303–333

Jiao W-B, Schneeberger K (2017) The impact of third generation genomic technologies on plant genome assembly. Curr Opin Plant Biol 36:64–70

Jiao Y, Peluso P, Shi J, Liang T, Stitzer MC, Wang B et al (2017) Improved maize reference genome with single-molecule technologies. Nature 546(7659):524–527

Lei Z, Xu C, Jian W, Min L, Grob S, Feng C et al (2018) Improved *Brassica rapa* reference genome by single-molecule sequencing and chromosome conformation capture technologies. Hort Res 5(1):50

Lysak MA, Koch MA, Pecinka A, Schubert I (2005) Chromosome triplication found across the tribe Brassiceae. Genome Res 15(4):516–525

Michael TP, VanBuren R (2015) Progress, challenges and the future of crop genomes. Curr Opin Plant Biol 24:71–81

Michael TP, Jupe F, Bemm F, Motley ST, Sandoval JP, Lanz C et al (2018) High contiguity *Arabidopsis thaliana* genome assembly with a single nanopore flow cell. Nat Commun 9(1):541

Nagaharu U (1935) Genome analysis in *Brassica* with special reference to the experimental formation of *B. napus* and peculiar mode of fertilization. Jpn J Bot 7 (7):389–452

Liu S, Liu Y, Yang X, Tong C, Edwards D, Parkin IA et al (2014) The *Brassica oleracea* genome reveals the asymmetrical evolution of polyploid genomes. Nat Commun 5

Parkin IA, Koh C, Tang H, Robinson SJ, Kagale S, Clarke WE et al (2014) Transcriptome and methylome profiling reveals relics of genome dominance in the mesopolyploid *Brassica oleracea*. Genome Biol 15 (6):R77

Ross MG, Russ C, Costello M, Hollinger A, Lennon NJ, Hegarty R et al (2013) Characterizing and measuring bias in sequence data. Genome Biol 14(5):R51

Seung Chul S, Do Hwan A, Jin KS, Hyoungseok L, Tae-Jin O, Jong Eun L et al (2013) Advantages of single-molecule real-time sequencing in high-GC content genomes. PLoS One 8(7):e68824

Staňková H, Hastie AR, Chan S, Vrána J, Tulpová Z, Kubaláková M et al (2016) BioNano genome mapping of individual chromosomes supports physical mapping and sequence assembly in complex plant genomes. Plant Biotechnol J 14(7):1523–1531

Sun S, Zhou Y, Jian C, Shi J, Zhao H, Zhao H et al (2018) Extensive intraspecific gene order and gene structural variations between Mo17 and other maize genomes. Nat Genet 50(9):1289

Van Dijk E, Jaszczyszyn Y, Naquin D, Thermes C (2018) The third revolution in sequencing technology. Trends Genet 34(9):666–681

VanBuren R, Bryant D, Edger PP, Tang H, Burgess D, Challabathula D et al (2015) Single-molecule sequencing of the desiccation-tolerant grass *Oropetium thomaeum*. Nature 527(7579):508

Wang X, Wang H, Wang J, Sun R, Wu J, Liu S et al (2011) The genome of the mesopolyploid crop species *Brassica rapa*. Nat Genet 43(10):1035–1039

Xiong Z, Pires JC (2011) Karyotype and identification of all homoeologous chromosomes of allopolyploid *Brassica napus* and its diploid progenitors. Genetics 187(1):37–49

Composition and Organization of Major Repeat Components in the *Brassica oleracea* Genome

5

Nomar Espinosa Waminal, Sampath Perumal, Jonghoon Lee, Hyun Hee Kim, and Tae-Jin Yang

Abstract

The release of the *Brassica oleracea* draft genome sequence opens numerous opportunities to understand its genome structure and evolution. A 515-Mb (82% of the total genome) high-quality draft assembly was made up of 56% repetitive elements (REs). Although the contribution of REs to genome structures, organization and evolution is relatively poorly understood, advances in bioinformatics have enabled genome-wide quantification and downstream analyses of REs in plant genomes. This chapter provides an overview of the classification, abundance, and genomic organization of the major types of REs that make up the main repeat component in the *B. oleracea* genome. Eight elements namely, 5S and 45S nrDNA, centromeric and subtelomeric tandem repeats (CentBo1, CentBo2, BoSTRa/b, and BoSTRc), a centromeric retrotransposon (BoCRB), and a *Ty1/copia* LTR retrotransposon (BoCopia-1) were classified into this repeat component. Whole-genome shotgun (WGS) mapping and molecular cytogenetic analyses provided an in-depth view of the abundance and distribution of these repeats both in the in silico generated draft assembly and mitotic metaphase chromosomes. The information not only validates the abundance of repeat elements in draft genomes, but also provide an avenue for understanding overall genome structure.

5.1 Introduction

Advances in genome sequencing technologies such as second-generations sequencing (SGS) and third-generations sequencing (TGS) have enabled genome assembly of many important crops in unprecedented contiguity and accuracy. This, in turn, provided a wealth of opportunity to understand the genome structure and function, and to perform comparative studies and a plethora of additional downstream analyses (Schadt et al. 2010; Edwards and Batley 2010; Schatz et al. 2012; Gonzaga-Jauregui et al. 2012).

N. E. Waminal and S. Perumal—Equally contributed.

N. E. Waminal · S. Perumal · J. Lee · T.-J. Yang (✉)
Department of Agriculture, Forestry and Bioresources, Plant Genomics & Breeding Institute, College of Agriculture and Life Sciences, Seoul National University, Seoul 08826, Republic of Korea
e-mail: tjyang@snu.ac.kr

N. E. Waminal · H. H. Kim
Bioscience Institute, Department of Chemistry and Life Science, Sahmyook University, Seoul 01795, Republic of Korea

S. Perumal
Agriculture and Agri-Food Canada, 107 Science Place, Saskatoon, SK S7N 0X2, Canada

S. Liu et al. (eds.), *The Brassica oleracea Genome*, Compendium of Plant Genomes,
https://doi.org/10.1007/978-3-030-31005-9_5

Whole-genome sequencing has revealed that a significant proportion of sequenced plant genomes consist of repetitive elements (REs) and only 2–5% of a genome is coding for functional genes (Lisch 2012; Feschotte and Pritham 2007). REs, which include transposable elements (TEs) and tandem repeats (TRs), occupy about 10–90% of a genome (Vitte et al. 2014; Goerner-Potvin and Bourque 2018). Compared to TRs, TEs are expected to be a significant contributor to genome size expansion. However, studies have also shown that TRs can be an essential cause to increase in genome size (Barghini et al. 2014). Different types of TEs (Class 1—Retrotransposon and Class II—DNA transposons) and their superfamilies have been characterized in plants and were reviewed in Wicker et al. (2007).

Compared to the quantum advancement of sequencing technologies, progress in developing efficient and universal assembly method has been somewhat quite sluggish, particularly in plants (Hamilton and Buell 2012), mainly because of the challenging nature in piecing together whole-genome sequence (WGS) reads from highly repetitive elements (Claros et al. 2012). For example, most genome assemblies produced by SGS technologies and assembly algorithms contain numerous gaps because of the difficulty in resolving the correct order of the numerous duplicated sequences caused by whole-genome duplications and REs (Schatz et al. 2012). Consequently, the presence of repetitive sequences is of significant concern in obtaining high-resolution pseudo-molecule-level genome assembly, especially that existing assembly algorithms are yet insufficient in handling these complex sequences (Schatz et al. 2012; Alkan et al. 2011; Claros et al. 2012). Although current advances in sequencing long read lengths offered by TGS, such as Oxford Nanopore Sequencing Technologies (ONT), is providing hope to resolve megabase longTR regions, high error rate (up to 15%) still confounds assembly (Schatz et al. 2012; Schadt et al. 2010; Jain et al. 2018; Jiao and Schneeberger 2017). Moreover, due to the complex rearrangement of repetitive elements, which mainly resides at the heterochromatic regions such as centromere and pericentromere, even some model plants like rice and *Arabidopsis* do not have a completely characterized organization of these regions (Gao et al. 2015).

However, although repeat sequences pose a tremendous challenge in genome assembly (Michael and Jackson 2013; Wang et al. 2011), they provide tremendous information on chromosome and genome evolution, gene regulation, and various epigenetic control. (Biemont 2010; Biémont and Vieira 2006; Chadwick 2009; Mehrotra and Goyal 2014). Several recent studies have shown the significant roles of these repetitive elements in genomes, refuting the longstanding concept of their useless existence (Nowak 1994; Shapiro and von Sternberg 2005; Doolittle and Sapienza 1980; Chadwick 2009; Mehrotra et al. 2014). Thus, they warrant a thorough genome-wide analysis. Understanding their genomic distribution would complement genome assemblies and genomics research (Biemont 2010; Choi et al. 2014). Here we review the overall landscape of the *B. oleracea* main repetitive component comprising eight REs by exploring the abundance and distribution of these REs through WGS and fluorescence in situ hybridization (FISH) analyses genome.

5.2 Repeat Composition of the *B. oleracea* Genome

B. oleracea (C_o genome, $2n = 2x = 18$, 630 Mb) is a highly important vegetable crop with wide morphological diversity. Along with other *Brassica* species, it also serves as an important model for understanding plant genome evolution. About 515 Mb (82% of the genome) was assembled to high-quality scaffolds (Fig. 5.1) (Liu et al. 2014; Chalhoub et al. 2014). Of this fraction, 385 Mbp (61%) was anchored into pseudo-chromosomes. Annotation of the *B. oleracea* genome has revealed that a significant proportion (56%) of the assembly consists of repeat elements including TEs and TRs (Table 5.1). Most TEs ($\sim$97%) are included in the genome assembly; conversely, TRs and other sequences are represented only 1 and 2%,

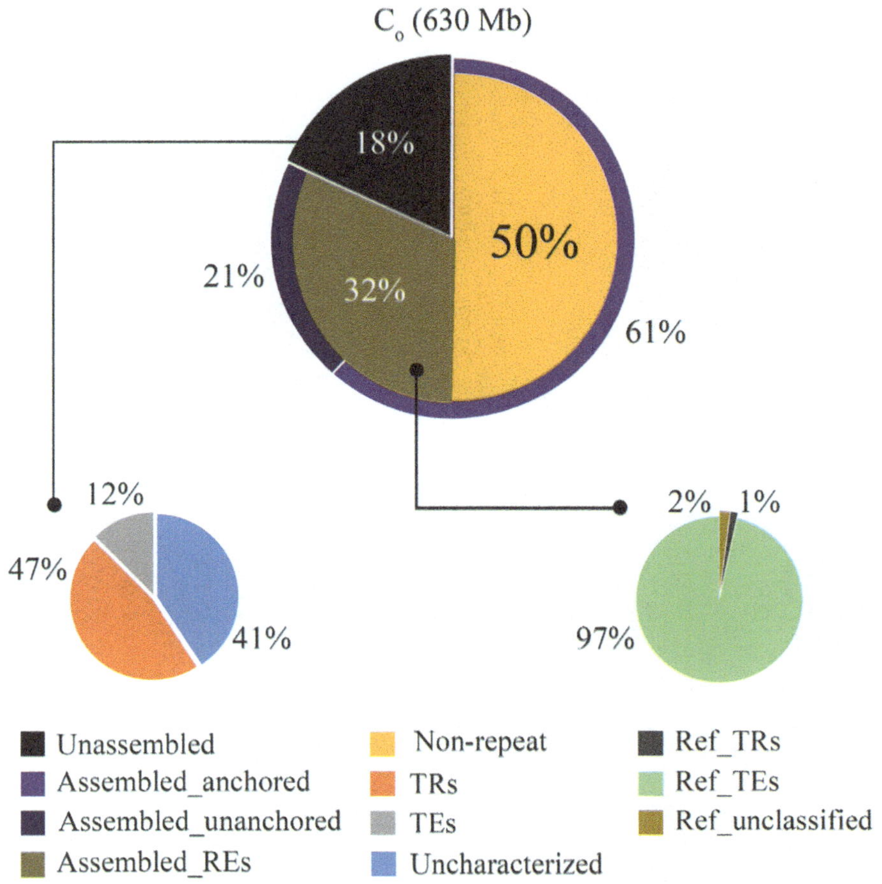

Fig. 5.1 Genomic proportions of the assembled and unassembled sequences identified in the *Brassica oleracea* genome. The large inner pie chart in the bigger circle represents the estimated repeat fraction (brown slice) and total non-repeat fractions (yellow slice) of *B. oleracea* reference genome assembly. The outer doughnut chart represents the assembly percentage (light and dark purple slices represent pseudochromosome-assembled and unassembled fractions) relative to the estimated genome size of 630 Mb as calculated by Liu et al. (2014). The black slice depicts the unassembled portion. Smaller pie charts at the bottom left and right summarizes the REs in the unassembled and assembled genome fractions, respectively

respectively. Among the assembled TEs, Class I had a higher proportion (40%) than Class II (13%) (Liu et al. 2014; Chalhoub et al. 2014). LTR retrotransposons represented the bulk of Class I, with the *Gypsy* type being more abundant followed by the *Copia* type (15 vs. 11%). On the other hand, *CACTA* elements (6%) were the most abundant among Class II TEs, followed by *Helitrons* (4%) and *Pong* (3%) (Liu et al. 2014).

B. oleracea has the largest genome among the three diploid genomes in the U's triangle (Nagaharu 1935); the other two being *B. rapa* (Ar-genome, $2n = 2x = 20$, 496 Mb) and *B. nigra* (B*n*, $2n = 2x = 16$, 570 Mb). Repeat expansion is the primary cause for an increase in genome size (Liu et al. 2014; Chalhoub et al. 2014). Also, in-depth genome-wide characterization of repeats and FISH analysis have enabled identification of a TE specific to the *B. oleracea* genome, BoCACTA, a high copy repeat from a CACTA family of Class II TE. BoCACTA was used to identify the Co sub-genome chromosomes in a *B. napus* genome (Perumal et al. 2017).

TEs without functional coding genes for their mobilization are called non-autonomous TEs or miniature TEs (mTEs) such as miniature inverted-repeat TEs (MITEs), short interspersed

Table 5.1 Composition of repeats in the genomes of *B. rapa*, *B. oleracea*, and *B. napus*

Repetitive elements			*B. rapa*			*B.oleracea*			*B. napus*		
Class/order	Superfamily	# of Families	Number	Length (Kb)	Percentage (%)	Number	Length (Kb)	Percentage (%)	Number	Length (Kb)	Percentage (%)
Class I—LTR transposons											
	Gypsy	229	33,164	16,637	6	91,058	78,454	15	113,420	96,798	15
	Copia	113	29,789	13,205	5	81,877	58,766	11	96,736	62,167	8
	Others/unclassified	13	50,600	14,746	5	90,105	36,553	7	124,306	47,103	7
	TRIM	72	4,257	1,860	1	5,200	3,432	1	7,183	3,639	1
				0			0				
	LINE	40	29,243	12,652	4	58,653	28,975	6	79,305	37,211	6
	SINE	154	10,279	1,955	1	15,039	3,029	1	20,276	4,030	1
Subtotal-Class I		**621**	**157332**	**61,056**	**21**	**341932**	**209,208**	**40**	**441,226**	**250,948**	**38**
Class II DNA transposon											
	DNA TE	143	35,560	13,103	5	90,133	51,157	10	111,137	54,018	8
	MITE	113	23,796	4,823	2	30,052	6,615	1	41,393	9,106	1
	Others/unclassified		6,435	1,990	1	12,859	6,053	1	16,511	7,109	1
	Helitron	63	4,505	1,636	1	9,392	3,447	1	12,751	4,426	1
Subtotal-Class II		**319**	**70,296**	**21,552**	**8**	**142,436**	**67,272**	**13**	**181,792**	**74,660**	**11**
Highcopy genes											
	nrDNA	2	14,341	4,196	1	19,479	5,590	1	27,025	7,548	1
Simple repeats		5	41,406	5,563	2	58,889	8,161	2	80,414	11,483	2
Total		**1890**	283,375	92,367	32	562,736	290,230	56	730,457	344,640	52

elements (SINEs), and terminal-repeat retrotransposons in miniature (TRIMs) (Sampath et al. 2015; Sampath and Yang 2014). Investigation of mTEs on Brassica genomes including *B. oleracea* revealed their abundance and wide distribution to the *B. oleracea* and related genomes. *B. oleracea* reference genome contains >3000 elements, representing about 3% of the genome (Table 5.1). The mTEs can influence gene structure and function upon insertion into a gene or its vicinity. For example, BraMi-1, a high copy stowaway superfamily MITE influenced the structural and functional divergence of genes in *B. rapa* (Sampath et al. 2013). Furthermore, an abundance of mTEs can serve as molecular markers for various genotyping and diversity studies (Murukarthick et al. 2014; Sampath et al. 2014).

Though knowledge on the potential genomic impacts of repeats has improved our understating of the *B. oleracea* genome, about 18% of the total estimated genome remains unassembled (Fig. 5.1). Contrary to what was observed in the assembled REs, most of the REs in the unassembled fraction was represented by TRs (47%), while TEs represented only about 12%. The rest of the elements were uncharacterized. These data corroborate the well-observed exclusion of TRs and TEs in plant genome sequence assemblies (Michael and Jackson 2013), and that TEs and TRs abound differentially in each assembled and unassembled fractions, as observed in *B. rapa* and *B. napus* (Waminal et al. 2015, 2018). Exploring these unassembled regions will give a better understanding of the landscape and dynamics of REs in the *B. oleracea* genome.

5.3 Exploring the Major Repeats in the Unassembled Fraction of the *B. oleracea* Genome

Estimation of genome proportion (GP) of REs using NGS data has been demonstrated successfully in various organisms (Macas et al. 2007; Hawkins et al. 2006; Macas et al. 2011; Piednoel et al. 2012; Maumus and Quesneville 2014). For instance, the use of SGS data has enabled identification of up to 48% of repeats in the 4300 Mbp of *Pisum sativum* genome (Macas et al. 2007). Here, we analyzed the unassembled fraction of the *B. oleracea* genome through a similar approach, using 1 × haploid equivalent WGS data generated on the Illumina HiSeq 2000 platform. We mapped these WGS reads onto the major repeats that were characterized in previous studies(Perumal et al. 2017). These include two LTR retrotransposons—acentromeric LTR retrotransposon, BoCRB (Lim et al. 2007), and a newly characterized LTR retrotransposon, BoCopia-1 (Perumal et al. 2017)—and six satellite repeats including 5S and 45S nrDNA (Waminal et al. 2015); centromeric TRs, CentBo1 and CentBo2 (Lim et al. 2005, 2007); and subtelomeric repeats, BoSTRa/b and BoSTRc (Koo et al. 2011) (Table 5.2).

The eight REs cover only 2.9 Mbp in the assembled fraction representing <0.5% of the 515 Mb *B. oleracea* reference genome (Table 5.2). Among the eight REs, BoSTRc had the highest representation in both copy numbers and GP. Conversely, these REs represented 10.5% of the genome upon mapping of 1x WGS read. Overall, the current assembly included only about 5% of the REs present in the fraction captured by WGS mapping (Table 5.2). The values for 1x WGS reads analysis denote average values for each repeat in 44 *B. oleracea* accessions (Perumal et al. 2017).

A similar analysis has also been done on the *B. rapa* genome showing 19% of the genome has been occupied with the related repeat family (Perumal et al. 2017). Compared to *B. rapa*, less proportion of the nine major repeats were represented in *B. oleracea* ($\sim$11%). However, in both the A_r and C_o genomes, CentBr/Bo1 was the most abundant repeat, comprising about 7% and 4%, respectively (Table 5.3), a comparatively less copy number in the C_o genome than that in the A_r genome, coherent to previous findings (Koo et al. 2011; Lim et al. 2007; Waminal et al., in press). While CentBo2 was the second most abundant repeat in the C_o genome (3%), 45S nrDNA was the second in the A_r genome (6%). The 45S and 5S nrDNA copies are

Table 5.2 Quantification of major repeats present in the reference genome and 1x WGS of *B. oleracea*

Repeat element	Unit length (bp)	Amounts in *B. oleracea* reference genome							Estimated amounts in *B. oleracea* 1x wgs			GP by FISH (%)	Uncovered portion (%)[c]
		Pseudomolecule (385 Mb)		Unanchored scaffolds (130 Mbp)		Total (515 Mbp)		GP[b] (%)	Copy (±SD)	Kb(±SD)	GP[b] (%)		
		Copy	Kb	Copy	Kb	Copy	Kb						
5S nrDNA	503	55	26.9	88	43.2	143	70.1	0.013	1,286 (±511)	647 (±257)	0.11	0.75	10.73
45S nrDNA	7,586	1	5.1	0	0	1	5.1	0.001	1,072 (±608)	8,136 (±4,611)	1.46	1.63	0.06
CentBo1	177	460	74.8	743	120.9	1,203	195.6	0.040	114,077 (±2,7198)	20,192 (±4,814)	3.58	7.3	0.96
CentBo2	177	510	83.2	1,414	233.9	1,924	317.1	0.059	89,827 (±26,227)	15,899 (±4,642)	2.88	6.17	1.93
BoSTRa/b	354	820	258.1	691	218.9	1,511	477	0.088	3,810 (±1,043)	1,354 (±369)	0.25	2.55	33.97
BoSTRc	351	3,114	1,040.5	2,072	694.4	5,186	1,734.9	0.321	21,067 (±5,708)	7,394 (±2,004)	1.34	4.67	22.65
BoCRB	6,156	2	11.6	0	0	2	11.6	0.002	486 (±105)	2,995 (±646)	0.54	2.98	0.38
BoCopia-1	6,681	10	58.9	5	28.7	15	87.6	0.016	298 (±49)	1,988 (±329)	0.36	1.75	4.26
Total		4,972	1,559.1	5,013	1,340	9,985	2,899	0.54	231,923 (±1,449)	58,605 (±17,671)	10.52	21.38	4.83

[a]Repeats were estimated based on the *in silico* read mapping analysis on 44 *Brassica oleracea* accessions and the average values were represented to 1x wgs coverage (630 Mb).wgs of B. oleracea was used from a previous study (Perumal S et al. 2017)

[b]Genome proportion (GP): (Total amounts in Kb/630,000 Kb) × 100

[c]Percentage of repeat amounts uncovered in reference genome compared to the estimated amounts in 1x wgs

Kb: amounts in kilobase pair, SD: standard deviation

Table 5.3 Comparison of repeat abundance between *B. rapa* and *B. oleracea*

Element ID	Element size (bp)	*B. rapa*						*B. oleracea*					
		Reference genome (283 Mbp)			Estimated amounts in1x wgs (529 Mbp)[a]			Reference genome (515 Mbp)			Estimated amounts in1x wgs (630 Mbp)[b]		
		Copy	Kb	GP (%)	Copy	Kb	GP (%)	Copy	Kb	GP (%)	Copy	Kb	GP (%)
5S nrDNA	501	17	8.4	0.002	5,588	2,800	0.5	143	70.1	0.011	1,286	647	0.11
45S nrDNA	7,456	1	4.2	0.001	4,395	32,766	6.2	1	5.1	0.001	1,072	8,136	1.46
CENTBr1/Bo1	176	145	24.3	0.005	197,157	34,700	6.6	1,203	195.6	0.031	114,077	20,192	3.58
CENTBr2/Bo2	176	215	35.8	0.007	40,312	7,095	1.3	1,924	317.1	0.05	89,827	15,899	2.88
BrSTRa/BoSTRa/b	352	1,323	435.3	0.082	14,579	5,137	1	1,511	477.0	0.076	3,810	1354	0.25
BrSTRb/BoSTRc	352	178	57.6	0.011	809	285	0.1	5,186	1,734.9	0.275	21,067	7,394	1.34
CRB	5,908	1	5.9	0.001	694	4,098	0.8	2	11.6	0.002	486	2,995	0.54
pCRBr	8,395	0	0	0	1,203	10,426	2	0	0	0	46	391	0.07
BoCopia-1	6,711	1	5.2	0.001	37	251	0	15	87.6	0.014	298	1,988	0.36
Total		1,881	576.7	0.109	264,961	99,788	18.9	9,985	2,899.1	0.46	231,969	58,996	10.59

[a,b]wgs of *B. rapa* and *B. oleracea* were used from a previous study (Perumal et al. 2017)

much less than those found in *B. rapa,* reflecting the fewer nrDNA loci of *B. oleracea* than that of *B. rapa*. These data are consistent with previous in silico and FISH results (Xiong and Pires 2011; Waminal et al., in press).

Two classes of subtelomeric TRs have been reported in Brassica, STRa and STRb, which share 87% sequence identity (Koo et al. 2011). STRa in *B. rapa* (BrSTRa) is homologous to STRa/b in *B. oleracea* (BoSTRa/b) while BrSTRb is to BoSTRc (Waminal et al. 2018). BoSTRc is more abundant than BoSTRa/b, contrary to the pattern observed in *B. rapa* (i.e., BrSTRa is more abundant than BrSTRb), suggesting a genome-specific amplification/elimination of different STR variants after species divergence (Waminal et al. 2018).

Between the two TEs included in this survey, BoCRB was more abundant than BoCopia-1 (Table 5.2). However, more BoCopia-1 was included in the reference genome than BoCRB. This case is most likely caused by the variation of the genomic distribution pattern of these two repeats; BoCRB are rich in condensed centromeric regions (Lim et al. 2007; Liu et al. 2014) but the BoCopia-1 is widely dispersed in the genome (Fig. 5.2a, b). Centromeres are reputed to be tough to assemble due to the highly repetitive nature of this region (Nagaki et al. 2004). Thus, many assembled genomes include only a few centromeric repeat elements such as CentBo and centromeric retrotransposon of Brassica (CRB)regardless of their copies in a genome (Nagaki et al. 2004; Melters et al. 2013).

5.4 Cytogenetic Mapping of the Major Repetitive Elements

With FISH, newly identified TRs were successfully mapped on metaphase chromosomes, painting an actual picture of their genomic distribution and abundance. This approach is particularly practical in studying the abundance and distribution of repetitive elements to support genome assemblies, mainly because in silico computational analysis did not always accurately report the proportion of repeats in published genomes (Macas et al. 2007; Schatz et al. 2012).

FISH data have enabled confirmation of chromosomal segment inversions (Huang et al. 2009; van der Knaap et al. 2004; Cabo et al. 2014), delimited centromeric regions (Wolfgruber et al. 2009; Lee et al. 2005), mapping of transposon distribution (Neumann et al. 2011; Kim et al. 2018) as well as other repetitive elements (Macas et al. 2007; Lamb et al. 2007b; Suzuki et al. 2012), and even detected single genes (Lamb et al. 2007a; Khrustaleva and Kik 2001; Danilova et al. 2014) as well as transgene insertion (Santos et al. 2006).

In this work, FISH data have provided a better estimate and visualization of the genomic abundance and distribution of each repeat element. Whereas about 10% of the total repetitive elements were captured using 1x WGS analysis, FISH revealed about 21% of all the repetitive elements in the genome (Table 5.2). However, we consider the FISH-based estimates to be an overestimate because FISH signals were measured based only on two-dimensional signal area contrary to the three-dimensional structure of a chromosome, and also due to a wider area that is produced by fluorescence signals than the actual hybridization loci.

There are one and two 5S and 45S nrDNA loci, respectively, in *B. oleracea* (Fig. 5.2a–c) (Liu et al. 2014), fewer compared with *B. rapa*, which has three and five (Xiong and Pires 2011; Lim et al. 2005; Koo et al. 2011). With relatively similar signal dimensions, both the fewer loci number and larger genome size of *B. oleracea* explain the much-reduced genome proportions of ribosomal DNAs in *B. oleracea* than in *B. rapa* (Waminal et al. 2018).

In both 1x WGS and FISH analyses, CentBo1 appeared to be the most abundant repeat followed by CentBo2, while 5S nrDNA was the least (Table 5.2). Collectively, centromeric repeats were the most abundant TR in the *B. oleracea* genome followed by the subtelomeric repeats. However, an observed discrepancy in the order of abundance of repeats when comparing

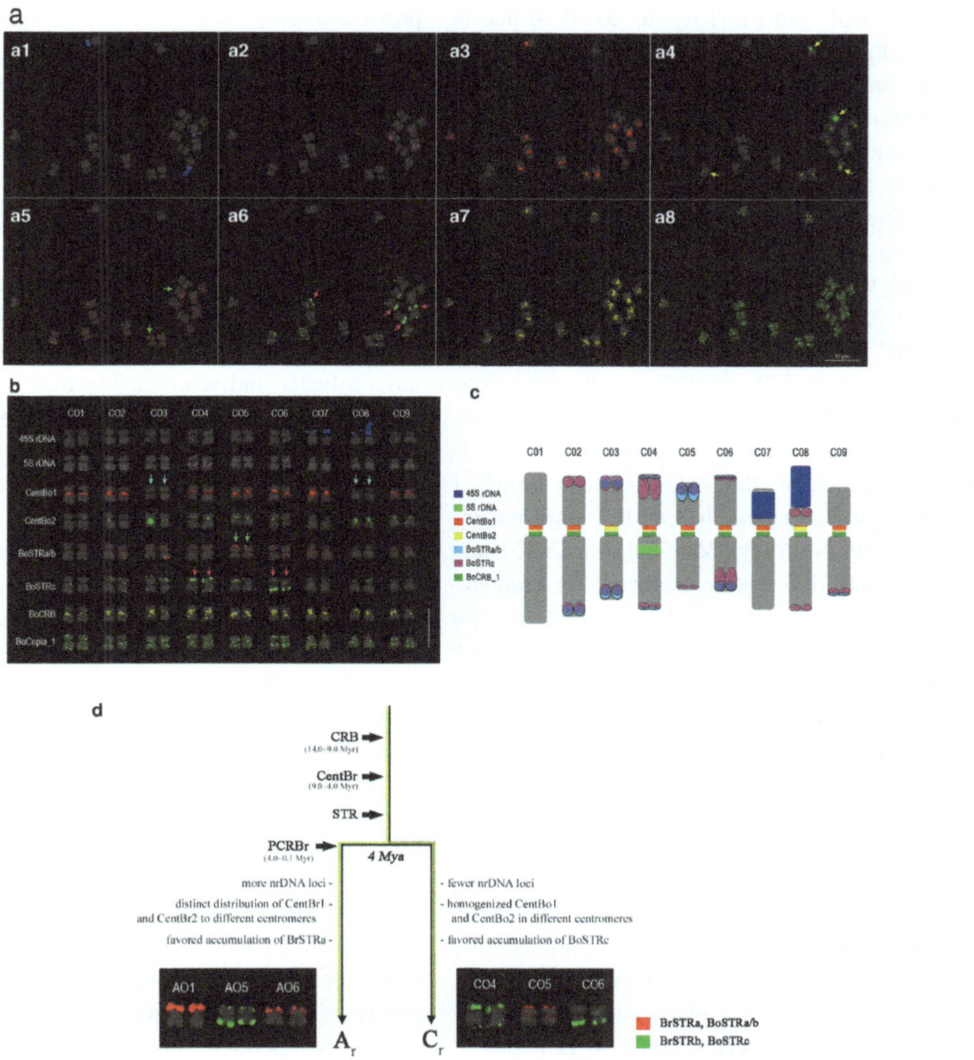

Fig. 5.2 Genomic distribution of major DNA TRs in *Brassica oleracea* ($2n = 2x = 18$, CC) through FISH analysis. **a** FISH analysis of a1 45S nrDNA, (a2) 5S nrDNA, (a3) CentBo1, (a4) CentBo2, (a5) BoSTRa/b), (a6) BoSTRc, (a7) BoCRB, and (a8) BoCopia-1. Green and pink arrows indicate major BoSTRa/b and BoSTRc signals, respectively, totaling three pairs of major BoSTR signals (as observed by Koo et al. 2011). Our more sensitive FISH method was able to detect minor STR loci not observed in the previous study. The TRs covered about 28% of the *B. oleracea* genome based on FISH estimation. **b** Karyogram of *Brassica oleracea* based on the distribution of major DNA TRs. Blue-green arrows indicate chromosomes with very weak CentBo1 while with strong CentBo2 signals. Most other chromosomes have colocalization of CentBo1 and CentBo2. Green and pink arrows indicate major signals of BoSTRa/b and BoSTRc, respectively, which totaled three major signals. Bar = 10 μm. **c** Karyotype idiogram of *Brassica oleracea*. Centromeric bars represent signal hybridization intensities represented arbitrarily. **d** Evolutionary diagram of major repetitive DNA in *B. oleracea*, showing a period of lineage divergence. PCRBr is a *B. rapa*-specific pericentromeric retrotransposon (Lim et al. 2007). Distinguishing features between genomes A and C after divergence is given. Homologous chromosomes in A_r and C_o genomes bearing major STR loci are shown in the inset images. The genome-specific amplification/elimination of STRs are demonstrated by the differential abundance of STR variants between the two subgenomes. Bars in A and B = 10 μm

both 1x WGS and FISH results could be due to an insufficient capture of repeats 1x WGS or overestimation using FISH.

In *B. rapa*, CentBr1 and CentBr2 show about 85% sequence similarity and are distinctly distributed to eight and two chromosome pairs, respectively (Lim et al. 2007). However, in *B. oleracea*, aside from the two homologues that showed intense CentBo2 signals, there is substantial colocalization of both CentBo1 and CentBo2 in other chromosomes with varying loci intensities (Fig. 5.2b, c), (Lim et al. 2007; Koo et al. 2011; Liu et al. 2014). These colocalizations are consistent with the findings that there is little copy number difference between CentBo1 and CentBo2 compared to CentBr1 and CentBr2 in the 1x WGS survey (Waminal et al. 2015), and also indicate different centromeric TR homogenization rate between *B. oleracea* and *B. rapa* genomes as well as among centromeres within each genome, as observed in some Brassicaceae species (Hall et al. 2005).

In the *B. oleracea* lineage (A_r and C_o genomes) CentBr/Bo is intermingled with the CRB. Although the CRB is common to the three primary Brassica lineages A_r, B_o, and C_o genomes, CentBr/Bo is present only in the *B. oleracea* lineage (Fig. 5.2b–d) (Lim et al. 2007). Additionally, BoCopia-1 which has preferentially amplified in the C_o genome hybridized to intercalary regions of *B. oleracea* chromosomes (Fig. 5.2b) but undetected in the A_r genome.

Previous analysis identified two variants of BoSTRs, but these variants were not differentially mapped on metaphase chromosomes (Koo et al. 2011). Our FISH data show a differential chromosomal distribution of these two subclades (Fig. 5.2b, c). Intense BoSTRa/b signals were observed at the subtelomeres of the short arm of chromosome C05 while chromosomes C04 and C06 have intense BoSTRc signals at the short arm and long arm, respectively. Moreover, aside from these three intense STR signals, which could have been the only signals observed in the previous report (Koo et al. 2011), several minor BoSTR loci, especially BoSTRc, were found at both arms of several chromosomes. More BoSTRc signals were observed compared with BoSTRa/b (Fig. 5.2), corroborating results from 1x WGS survey. Altogether, the satellite repeat distribution in *B. oleracea* further supports the observation that satellite repeats often accumulate at centromeric and subtelomeric regions (Charlesworth et al. 1994; Sharma and Raina 2005). Cytogenetic mapping of these repeats gave a better and reproducible method in identifying individual *B.* oleracea chromosomes (Table 5.4). A summary of the cytogenetic distribution of *B. oleracea* repeats is presented in Fig. 5.2c.

As has been done in *Pisum sativum* (Macas et al. 2007) and *B. rapa* (Waminal et al. 2015), analyzing low-coverage NGS data of plant genomes is practical for quantifying highly repetitive genomic components which are often omitted during pseudomolecule-level assembly. Through this approach, analysis of repetitive elements using WGS data could provide an enormous amount of information regarding genome structure and evolution, especially when comparative studies are performed with closely and distantly related species. This approach may also enable us to understand genome organization of plants with huge genomes such as *Allium* before the availability of genome assembly (Jakse et al. 2008). REs in the repetitive component of a plant species can be analyzed using low-coverage WGS before the completion of a draft genome assembly.

5.5 Functions and Evolutionary Implications of Repetitive Elements

Repetitive DNAs are distributed as either dispersed or clustered in eukaryotic genomes. TEs comprise a vast proportion of dispersed repeats, while TRs comprise the clustered repeats, which

Table 5.4 Summary of chromosome features based on the distribution of major TRs

Chr. no.	Features
C01	Strong CentBo1 signal, centromeric localization of BoSTRa/b, weak BoSTRc at the centromere
C02	Strong CentBo1 signal, centromeric and telomeric of long arm localization of BoSTRa/b, centromeric weak BoSTRc and BoSTRc telomeric on both arms
C03	Very weak CentBo1 while strong CentBo2, telomeric at both arms and centromeric BoSTRa/b and telomeric BoSTRc at both arms
C04	With 5S nrDNA at the pericentromeric area of the long arm, with CentBo1 and CentBo2, telomeric at both arms and centromeric BoSTRa/b, telomeric BoSTRc at both arms with the signal on short arm being one of two major BoSTRc signals
C05	CentBo1 and CentBo2 present, BoSTRa/b centromeric and telomeric. BoSTRa/b on short arm is the only major BoSTRa/b signal, BoSTRc telomeric and weak
C06	With CentBo1 and weak CentBo2, Centromeric and telomeric BoSTRa/b on both arms, weak telomeric on the short arm but major BoSTRc signal on the long arm
C07	45S nrDNA on short arm, with CentBo1 but very weak CentBo2, centromeric BoSTRa/b, very weak BoSTRc
C08	45S nrDNA on short arm, very weak CentBo1 while strong CentBo2, weak centromeric BoSTRa/b, weak BoSTRc
C09	With CentBo1 and Centbo2, Centromeric and very weak telomeric BoSTRa/b on the long arm, weak telomeric BoSTRc on the long arm

include arrays of almost identical units in most plant genomes (Sharma and Raina 2005). Differences in the number of REs is mainly responsible for variation in plant genome sizes (Wei et al. 2013; Sharma and Raina 2005; Bennett and Leitch 2011), a phenomenon that is commonly referred to as the C-value enigma (Pagel and Johnstone 1992; Macas et al. 2007; Hardman 1986; Schulman and Wicker 2013). Recent evidence supports the important roles of these repeats for proper genome functions which relates to adaption and evolution (Nowak 1994; Pardue and DeBaryshe 2003; Biémont and Vieira 2006). TEs play significant roles in chromosome integrity by protecting chromosomes through a telomerase-independent fashion, and by maintaining centromeres, which represent a substantial role in chromosome segregation (Biemont 2010; Wolfgruber et al. 2009; Goodier et al. 2012). Additionally, many studies have shown the roles of TEs in phenotype variation and adaptive evolution in many organisms (Meagher and Vassiliadis 2005; Lisch 2013).

Satellite DNAs of 140–180 bp units and centromere-specific retrotransposons (CR) mostly comprise centromeres and pericentromeres of plants (Neumann et al. 2011;, Jiang et al. 2003; Jiang 2013; Talbert and Henikoff 2010; Sharma and Raina 2005). The centromeric TR arrays can extend to several megabase pairs and are often interrupted by a complex insertion of CRs, forming a complex nested pattern while playing a significant role in centromere function and evolution (Lim et al. 2007; Jiang et al. 2003; Wei et al. 2013). Association of the modified histone H3 (CENH3), which is a hallmark of active centromeres, with these TRs and CRs further suggest their active involvement in functional centromeres (Neumann et al. 2011; Jiang 2013).

Several findings have been presented to elucidate the rapid evolution of centromeric tandem arrays across different centromeres within a species. Gene conversion, unequal crossover, and repeat transposition are few of the models that attempt to explain the homogenization and amplification of these arrays between homologous chromosomes and non-homologous chromosomes and even between sister chromatids (Walsh 1987; Hall et al. 2005; Dover 1982). Higher-order repeat units often result from unequal crossovers. These higher-order repeats consist of several repeat units that have accumulated internal variations, and array lengths often vary as well (Talbert and Henikoff 2010). Other mechanisms like repeat transposition and gene conversion may amplify satellite arrays causing dispersion into other genomic regions (Hall et al. 2005; Dover 1986).

CRB is a common centromeric component of the A_r, B_n, and C_o genomes. However, the absence of CentBr/Bo FISH hybridization signals in the B_n genome implies the earlier divergence of the B_n from the A_r and C_o genomes (Fig. 5.2d; Lim et al. 2007; Arias et al. 2014). The same pattern was shown by FISH analysis of Brassica STR which showed genome-specific evolution of these subtelomeric repeats. The BnSTR from *B. nigra* did not hybridize to either the A_r or C_o genomes, and BrSTR from the A_r genome did not hybridize to either the B_n or C_o genome, although BoSTR from the C_o genome hybridized to both the A_r and C_o genomes (Koo et al. 2011). Moreover, the higher sequence similarity of the same STR subclades between species than that of between two subclades within a species (i.e., BoSTRa/b is more similar to BrSTRa than BoSTRa/b is to BoSTRc) suggests that divergence of the two subclades in the parental genomes of the Oleracea lineage happened before speciation took place (Fig. 5.2d; Liu et al. 2014).

The family Brassicaceae ancestral karyotype consisted of eight chromosomes which underwent whole-genome triplication and subsequent lineage-specific rearrangements resulting to nine chromosomes of *B. oleracea* which are generally longer than the *B. rapa* chromosomes (Liu et al. 2014). Moreover, genomic rearrangements may only lead to species divergence once a reproductive barrier is established (Sobel and Streisfeld 2015). However, these may not be sufficient enough to drive speciation but instead create distinct characteristics among populations adequate to be categorized as subspecies (Lin et al. 2012). *B. oleracea* subspecies are estimated to have diverged from each other around 1.44 Mya (Arias et al. 2014), and different subspecies or morphotype harbor different amount of repeat (Perumal et al. 2017); comparative cytogenetic mapping of these repetitive elements among the *B. oleracea* subspecies may provide a deeper insight of the on-going genome dynamics of recently diverged taxa.

5.6 Conclusions

TEs are important players not only in genome size evolution but also in the function of genes. Thus, they play an important role in the evolution of a genome; hence, adaptation of a species. Along with the three important diploid Brassica crops, *B. oleracea* provides an excellent platform in understanding the RE dynamics in Brassica because of the diversity, abundance, and differential chromosomal distribution of the REs in the main repetitive genomic component. Exploring the complete genome information including the repeat component is vital for the basic understanding of the *B. oleracea* genome. As we described in this chapter, the unassembled fraction of the *B. oleracea* genome is mainly represented by few classes of repeats, highlighting the need for methods to effectively assemble those repeat regions. Improvement of long-read genome sequencing technologies such as ONT and Pac-Bio sequencing will help to assemble the hidden space more efficiently, which would lead to a better understanding of the genome structure, organization, and evolution. Furthermore, it is still not clear how TEs are impacting the Brassica genome evolution, functional divergence, and contribution to the genome diversity. Future studies directed at answering these questions would dramatically enhance our view of TEs in general.

References

Alkan C, Sajjadian S, Eichler EE (2011) Limitations of next-generation genome sequence assembly. Nat Methods 8:61–65

Arias T, Beilstein MA, Tang M, McKain MR, Pires JC (2014) Diversification times among Brassica (Brassicaceae) crops suggest hybrid formation after 20 million years of divergence. Am J Bot 101:86–91

Barghini E, Natali L, Cossu RM, Giordani T, Pindo M, Cattonaro F, Scalabrin S, Velasco R, Morgante M, Cavallini A (2014) The peculiar landscape of repetitive sequences in the olive (Olea europaea L.) genome. Genome Biol Evol 6:776–791

Bennett MD, Leitch IJ (2011) Nuclear DNA amounts in angiosperms: targets, trends and tomorrow. Ann Bot 107:467–590

Biemont C (2010) A brief history of the status of transposable elements: from junk DNA to major players in evolution. Genetics 186:1085–1093

Biémont C, Vieira C (2006) Genetics: junk DNA as an evolutionary force. Nature 443:521–524

Cabo S, Carvalho A, Martin A, Lima-Brito J (2014) Structural rearrangements detected in newly-formed hexaploid tritordeum after three sequential FISH experiments with repetitive DNA sequences. J Genet 93:183–188

Chadwick BP (2009) Macrosatellite epigenetics: the two faces of DXZ4 and D4Z4. Chromosoma 118:675–81

Chalhoub B, Denoeud F, Liu S, Parkin IAP, Tang H, Wang X, Chiquet J, Belcram H, Tong C, Samans B, Corréa M, da Silva C, Just J, Falentin C, Koh CS, le Clainche I, Bernard M, Bento P, Noel B, Labadie K, Alberti A, Charles M, Arnaud D, Guo H, Daviaud C, Alamery S, Jabbari K, Zhao M, Edger PP, Chelaifa H, Tack D, Lassalle G, Mestiri I, Schnel N, le Paslier M-C, Fan G, Renault V, Bayer PE, Golicz AA, Manoli S, Lee T-H, Thi VHD, Chalabi S, Hu Q, Fan C, Tollenaere R, Lu Y, Battail C, Shen J, Sidebottom CHD, Wang X, Canaguier A, Chauveau A, Bérard A, Deniot G, Guan M, Liu Z, Sun F, Lim YP, Lyons E, Town CD, Bancroft I, Wang X, Meng J, Ma J, Pires JC, King GJ, Brunel D, Delourme R, Renard M, Aury J-M, Adams KL, Batley J, Snowdon RJ, Tost J, Edwards D, Zhou Y, Hua W, Sharpe AG, Paterson AH, Guan C, Wincker P (2014) Early allopolyploid evolution in the post-Neolithic Brassica napus oilseed genome. Science 345:950–953

Charlesworth B, Sniegowski P, Stephan W (1994) The evolutionary dynamics of repetitive DNA in eukaryotes. Nature 371:215–220

Choi HI, Waminal NE, Park HM, Kim NH, Choi BS, Park M, Choi D, Lim YP, Kwon SJ, Park BS, Kim HH, Yang TJ (2014) Major repeat components covering one-third of the ginseng (*Panax ginseng* C. A. Meyer) genome and evidence for allotetraploidy. Plant J 77:906–916

Claros MG, Bautista R, Guerrero-Fernandez D, Benzerki H, Seoane P, Fernandez-Pozo N (2012) Why assembling plant genome sequences is so challenging. Biology (Basel) 1:439–459

Danilova TV, Friebe B, Gill BS (2014) Development of a wheat single gene FISH map for analyzing homoeologous relationship and chromosomal rearrangements within the Triticeae. Theor Appl Genet 127:715–730

Doolittle WF, Sapienza C (1980) Selfish genes, the phenotype paradigm and genome evolution. Nature 284:601–603

Dover G (1982) Molecular drive: a cohesive mode of species evolution. Nature 299:111–117

Dover GA (1986) Molecular drive in multigene families: how biological novelties arise, spread and are assimilated. Trends Genetics 2:159–165

Edwards D, Batley J (2010) Plant genome sequencing: applications for crop improvement. Plant Biotechnol J 8:2–9

Feschotte C, Pritham EJ (2007) DNA transposons and the evolution of eukaryotic genomes. Annu Rev Genet 41:331–368

Gao D, Jiang N, Wing RA, Jiang J, Jackson SA (2015) Transposons play an important role in the evolution and diversification of centromeres among closely related species. Front Plant Science 6

Goerner-Potvin P, Bourque G (2018) Computational tools to unmask transposable elements. Nat Rev Genet 19:688–704

Gonzaga-Jauregui C, Lupski JR, Gibbs RA (2012) Human genome sequencing in health and disease. Ann Rev Med 63:35–61

Goodier JL, Cheung LE, Kazazian HH Jr (2012) MOV10 RNA Helicase is a potent inhibitor of retrotransposition in cells. PLoS Genet 8:e1002941

Hall SE, Luo S, Hall AE, Preuss D (2005) Differential rates of local and global homogenization in centromere satellites from arabidopsis relatives. Genetics 170:1913–1927

Hamilton JP, Buell CR (2012) Advances in plant genome sequencing. Plant J 70:177–190

Hardman N (1986) Structure and function of repetitive DNA in eukaryotes. Biochem J 234:1–11

Hawkins JS, Kim H, Nason JD, Wing RA, Wendel JF (2006) Differential lineage-specific amplification of transposable elements is responsible for genome size variation in Gossypium. Genome Res 16:1252–1261

Huang S, Li R, Zhang Z, Li L, Gu X, Fan W, Lucas WJ, Wang X, Xie B, Ni P, Ren Y, Zhu H, Li J, Lin K, Jin W, Fei Z, Li G, Staub J, Kilian A, Van Der Vossen EA, Wu Y, Guo J, He J, Jia Z, Tian G, Lu Y, Ruan J, Qian W, Wang M, Huang Q, Li B, Xuan Z, Cao J, Asan Wu Z, Zhang J, Cai Q, Bai Y, Zhao B, Han Y, Li Y, Li X, Wang S, Shi Q, Liu S, Cho WK, Kim JY, Xu Y, Heller-Uszynska K, Miao H, Cheng Z, Zhang S, Wu J, Yang Y, Kang H, Li M, Liang H, Ren X, Shi Z, Wen M, Jian M, Yang H, Zhang G, Yang Z, Chen R, Ma L, Liu H, Zhou Y, Zhao J, Fang X, Fang L, Liu D, Zheng H, Zhang Y, Qin N,

Li Z, Yang G, Yang S, Bolund L, Kristiansen K, Li S, Zhang X, Wang J, Sun R, Zhang B, Jiang S, Du Y (2009) The genome of the cucumber, *Cucumis sativus* L. Nat Gene 41:1275–1281

Jain M, Koren S, Miga KH, Quick J, Rand AC, Sasani TA, Tyson JR, Beggs AD, Dilthey AT, Fiddes IT, Malla S, Marriott H, Nieto T, O'Grady J, Olsen HE, Pedersen BS, Rhie A, Richardson H, Quinlan AR, Snutch TP, Tee L, Paten B, Phillippy AM, Simpson JT, Loman NJ, Loose M (2018) Nanopore sequencing and assembly of a human genome with ultra-long reads. Nat Biotechnol 36:338–345

Jakse J, Meyer JD, Suzuki G, McCallum J, Cheung F, Town CD, Havey MJ (2008) Pilot sequencing of onion genomic DNA reveals fragments of transposable elements, low gene densities, and significant gene enrichment after methyl filtration. Mol Genet Genom 280:287–292

Jiang J (2013) Centromere evolution. Plant Centromere Biology. Wiley-Blackwell

Jiang J, Birchler JA, Parrott WA, Kelly Dawe R (2003) A molecular view of plant centromeres. Trends Plant Sci 8:570–575

Jiao W-B, Schneeberger K (2017) The impact of third generation genomic technologies on plant genome assembly. Curr Opin Plant Biol 36:64–70

Khrustaleva LI, Kik C (2001) Localization of single-copy T-DNA insertion in transgenic shallots (*Allium cepa*) by using ultra-sensitive FISH with tyramide signal amplification. Plant J 25:699–707

Kim NH, Jayakodi M, Lee SC, Choi BS, Jang W, Lee J, Kim HH, Waminal NE, Lakshmanan M, van Nguyen B, Lee YS, Park HS, Koo HJ, Park JY, Perumal S, Joh HJ, Lee H, Kim J, Kim IS, Kim K, Koduru L, Kang KB, Sung SH, Yu Y, Park DS, Choi D, Seo E, Kim S, Kim YC, Hyun DY, Park YI, Kim C, Lee TH, Kim HU, Soh MS, Lee Y, In JG, Kim HS, Kim YM, Yang DC, Wing RA, Lee DY, Paterson AH, Yang TJ (2018) Genome and evolution of the shade-requiring medicinal herb Panax ginseng. Plant Biotechnol J 16:1904–1917

van der Knaap E, Sanyal A, Jackson SA, Tanksley SD (2004) High-resolution fine mapping and fluorescence in situ hybridization analysis of sun, a locus controlling tomato fruit shape, reveals a region of the tomato genome prone to DNA rearrangements. Genetics 168:2127–2140

Koo DH, Hong CP, Batley J, Chung YS, Edwards D, Bang JW, Hur Y, Lim YP (2011) Rapid divergence of repetitive DNAs in Brassica relatives. Genomics 97:173–185

Lamb JC, Danilova T, Bauer MJ, Meyer JM, Holland JJ, Jensen MD, Birchler JA (2007a) Single-gene detection and karyotyping using small-target fluorescence in situ hybridization on maize somatic chromosomes. Genetics 175:1047–1058

Lamb JC, Meyer JM, Corcoran B, Kato A, Han F, Birchler JA (2007b) Distinct chromosomal distributions of highly repetitive sequences in maize. Chromosome Res 15:33–49

Lee HR, Zhang W, Langdon T, Jin W, Yan H, Cheng Z, Jiang J (2005) Chromatin immunoprecipitation cloning reveals rapid evolutionary patterns of centromeric DNA in Oryza species. Proc Natl Acad Sci USA 102:11793–11798

Lim KB, de Jong H, Yang TJ, Park JY, Kwon SJ, Kim JS, Lim MH, Kim JA, Jin M, Jin YM, Kim SH, Lim YP, Bang JW, Kim HI, Park BS (2005) Characterization of rDNAs and tandem repeats in the heterochromatin of Brassica rapa. Mol Cells 19:436–444

Lim KB, Yang TJ, Hwang YJ, Kim JS, Park JY, Kwon SJ, Kim J, Choi BS, Lim MH, Jin M, Kim HI, de Jong H, Bancroft I, Lim Y, Park BS (2007) Characterization of the centromere and pericentromere retrotransposons in Brassica rapa and their distribution in related Brassica species. Plant J 49:173–183

Lin H, Xia P, Wing A, Zhang Q, Luo M (2012) Dynamic intra-japonica subspecies variation and resource application. Mol Plant 5:218–230

Lisch D (2012) How important are transposons for plant evolution? Nat Rev Gene 14:49

Lisch D (2013) How important are transposons for plant evolution? Nat Rev Genet 14:49–61

Liu S, Liu Y, Yang X, Tong C, Edwards D, Parkin IAP, Zhao M, Ma J, Yu J, Huang S, Wang X, Wang J, Lu K, Fang Z, Bancroft I, Yang T-J, Hu Q, Wang X, Yue Z, Li H, Yang L, Wu J, Zhou Q, Wang W, King GJ, Pires JC, Lu C, Wu Z, Sampath P, Wang Z, Guo H, Pan S, Yang L, Min J, Zhang D, Jin D, Li W, Belcram H, Tu J, Guan M, Qi C, Du D, Li J, Jiang L, Batley J, Sharpe AG, Park B-S, Ruperao P, Cheng F, Waminal NE, Huang Y, Dong C, Wang L, Li J, Hu Z, Zhuang M, Huang Y, Huang J, Shi J, Mei D, Liu J, Lee T-H, Wang J, Jin H, Li Z, Li X, Zhang J, Xiao L, Zhou Y, Liu Z, Liu X, Qin R, Tang X, Liu W, Wang Y, Zhang Y, Lee J, Kim HH, Denoeud F, Xu X, Liang X, Hua W, Wang X, Wang J, Chalhoub B, Paterson AH (2014) The Brassica oleracea genome reveals the asymmetrical evolution of polyploid genomes. Nat Commun 5:3930

Macas J, Kejnovsky E, Neumann P, Novak P, Koblizkova A, Vyskot B (2011) Next generation sequencing-based analysis of repetitive DNA in the model dioecious [corrected] plant Silene latifolia. PLoS One 6:e27335

Macas J, Neumann P, Navratilova A (2007) Repetitive DNA in the pea (*Pisum sativum* L.) genome: comprehensive characterization using 454 sequencing and comparison to soybean and *Medicago truncatula*. BMC Genomics 8:427

Maumus F, Quesneville H (2014) Deep investigation of Arabidopsis thaliana junk DNA reveals a continuum between repetitive elements and genomic dark matter. PLoS One 9:e94101

Meagher TR, Vassiliadis C (2005) Phenotypic impacts of repetitive DNA in flowering plants. New Phytol 168:71–80

Mehrotra S, Goel S, Raina SN, Rajpal VR (2014) Significance of satellite DNA revealed by conservation of a widespread repeat DNA sequence among angiosperms. Appl Biochem Biotechnol 173:1790–1801

Mehrotra S, Goyal V (2014) Repetitive sequences in plant nuclear dna: types, distribution, evolution and function. Genom Proteomics Bioinform 12:164–171

Melters DP, Bradnam KR, Young HA, Telis N, May MR, Ruby JG, Sebra R, Peluso P, Eid J, Rank D, Garcia JF, Derisi JL, Smith T, Tobias C, Ross-Ibarra J, Korf I, Chan SW (2013) Comparative analysis of tandem repeats from hundreds of species reveals unique insights into centromere evolution. Genome Biol 14: R10

Michael TP, Jackson S (2013) The first 50 plant genomes. Plant Genome 6

Murukarthick J, Sampath P, Lee SC, Choi B-S, Senthil N, Liu S, Yang T-J (2014) BrassicaTED—a public database for utilization of miniature transposable elements in Brassica species. BMC Res Notes 7:379

Nagaharu U (1935) Genome analysis in Brassica with special reference to the experimental formation of B. napus and peculiar mode of fertilization. Jap J Bot 7:389–452

Nagaki K, Cheng Z, Ouyang S, Talbert PB, Kim M, Jones KM, Henikoff S, Buell CR, Jiang J (2004) Sequencing of a rice centromere uncovers active genes. Nat Genet 36:138–145

Neumann P, Navratilova A, Koblizkova A, Kejnovsky E, Hribova E, Hobza R, Widmer A, Dolezel J, Macas J (2011) Plant centromeric retrotransposons: a structural and cytogenetic perspective. Mob DNA 2:4

Nowak R (1994) Mining treasures from 'junk DNA'. Science 263:608–610

Pagel M, Johnstone RA (1992) Variation across species in the size of the nuclear genome supports the junk-DNA explanation for the C-value paradox. Proc Royal Soc Lond. Ser B: Biol Sci 249:119–124

Pardue ML, Debaryshe PG (2003) Retrotransposons provide an evolutionarily robust non-telomerase mechanism to maintain telomeres. Ann Rev Genet 37:485–511

Perumal S, Waminal NE, Lee J, Lee J, Choi B-S, Kim HH, Grandbastien M-A, Yang T-J (2017) Elucidating the major hidden genomic components of the A, C, and AC genomes and their influence on Brassica evolution. Sci Rep 7:17986

Piednoel M, Aberer AJ, Schneeweiss GM, Macas J, Novak P, Gundlach H, Temsch EM, Renner SS (2012) Next-generation sequencing reveals the impact of repetitive DNA across phylogenetically closely related genomes of Orobanchaceae. Mol Biol Evol 29:3601–3611

Sampath P, Lee S-C, Lee J, Izzah NK, Choi B-S, Jin M, Park B-S, Yang T-J (2013) Characterization of a new high copy Stowaway family MITE, BRAMI-1 in Brassica genome. BMC Plant Biol 13:56

Sampath P, Yang T-J (2014) Comparative analysis of Cassandra TRIMs in three Brassicaceae genomes. Plant Genetic Resourc 12:S146–S150

Sampath P, Lee J, Cheng F, Wang X, Yang T-J (2015) Miniature Transposable Elements (mTEs): impacts and uses in the brassica genome. The Brassica rapa Genome. Springer

Sampath P, Murukarthick J, Izzah NK, Lee J, Choi H-I, Shirasawa K, Choi B-S, Liu S, Nou I-S, Yang T-J (2014) Genome-wide comparative analysis of 20 miniature inverted-repeat transposable element families in Brassica rapa and B. oleracea. PloS One 9: e94499

Santos AP, Wegel E, Allen GC, Thompson WF, Stoger E, Shaw P, Abranches R (2006) In situ methods to localize transgenes and transcripts in interphase nuclei: a tool for transgenic plant research. Plant Methods 2:18

Schadt EE, Turner S, Kasarskis A (2010) A window into third-generation sequencing. Hum Mol Genet 19: R227–R240

Schatz MC, Witkowski J, McCombie WR (2012) Current challenges in de novo plant genome sequencing and assembly. Genome Biol 13:243

Schulman AH, Wicker T (2013) A field guide to transposable elements. Plant transposons and genome dynamics in evolution. Wiley-Blackwell

Shapiro JA, von Sternberg R (2005) Why repetitive DNA is essential to genome function. Biol Rev Camb Philos Soc 80:227–250

Sharma S, Raina SN (2005) Organization and evolution of highly repeated satellite DNA sequences in plant chromosomes. Cytogenet Genome Res 109:15–26

Sobel JM, Streisfeld MA (2015) Strong premating reproductive isolation drives incipient speciation in Mimulus aurantiacus. Evolution 69:447–461

Suzuki G, Ogaki Y, Hokimoto N, Xiao L, Kikuchi-Taura A, Harada C, Okayama R, Tsuru A, Onishi M, Saito N, Do GS, Lee SH, Ito T, Kanno A, Yamamoto M, Mukai Y (2012) Random BAC FISH of monocot plants reveals differential distribution of repetitive DNA elements in small and large chromosome species. Plant Cell Rep 31:621–628

Talbert PB, Henikoff S (2010) Centromeres convert but don't cross. PLoS Biol 8:e1000326

Vitte C, Fustier M-A, Alix K, Tenaillon MI (2014) The bright side of transposons in crop evolution. Brief Funct Genom 13:276–295

Walsh JB (1987) Persistence of tandem arrays: implications for satellite and simple-sequence DNAs. Genetics 115:553–567

Waminal NE, Perumal S, Lim K-B, Park B-S, Kim HH, Yang T-J (2015) Genomic survey of the hidden components of the B. rapa genome. The Brassica rapa Genome. Springer

Waminal NE, Perumal S, Liu S, Chalhoub B, Kim HH, Yang T-J (2018) Quantity, distribution, and evolution of major repeats in Brassica napus. In: Liu S, Snowdon R, Chalhoub B (eds) The *Brassica napus* genome. Springer International Publishing, Cham

Wang X, Wang H, Wang J, Sun R, Wu J, Liu S, Bai Y, Mun JH, Bancroft I, Cheng F, Huang S, Li X, Hua W,

Wang J, Wang X, Freeling M, Pires JC, Paterson AH, Chalhoub B, Wang B, Hayward A, Sharpe AG, Park BS, Weisshaar B, Liu B, Li B, Liu B, Tong C, Song C, Duran C, Peng C, Geng C, Koh C, Lin C, Edwards D, Mu D, Shen D, Soumpourou E, Li F, Fraser F, Conant G, Lassalle G, King GJ, Bonnema G, Tang H, Wang H, Belcram H, Zhou H, Hirakawa H, Abe H, Guo H, Wang H, Jin H, Parkin IA, Batley J, Kim JS, Just J, Li J, Xu J, Deng J, Kim JA, Li J, Yu J, Meng J, Wang J, Min J, Poulain J, Wang J, Hatakeyama K, Wu K, Wang L, Fang L, Trick M, Links MG, Zhao M, Jin M, Ramchiary N, Drou N, Berkman PJ, Cai Q, Huang Q, Li R, Tabata S, Cheng S, Zhang S, Zhang S, Huang S, Sato S, Sun S, Kwon SJ, Choi SR, Lee TH, Fan W, Zhao X, Tan X, Xu X, Wang Y, Qiu Y, Yin Y, Li Y et al (2011) The genome of the mesopolyploid crop species Brassica rapa. Nat Genet 43:1035–1039

Wei L, Xiao M, An Z, Ma B, Mason AS, Qian W, Li J, Fu D (2013) New insights into nested long terminal repeat retrotransposons in Brassica species. Mol Plant 6:470–482

Wicker T, Sabot F, Hua-Van A, Bennetzen JL, Capy P, Chalhoub B, Flavell A, Leroy P, Morgante M, Panaud O (2007) A unified classification system for eukaryotic transposable elements. Nat Rev Genet 8

Wolfgruber TK, Sharma A, Schneider KL, Albert PS, Koo D-H, Shi J, Gao Z, Han F, Lee H, Xu R, Allison J, Birchler JA, Jiang J, Dawe RK, Presting GG (2009) Maize centromere structure and evolution: sequence analysis of Centromeres 2 and 5 reveals dynamic loci shaped primarily by retrotransposons. PLoS Genet 5:e1000743

Xiong ZY, Pires JC (2011) Karyotype and identification of all homoeologous chromosomes of allopolyploid *Brassica napus* and its diploid progenitors. Genetics 187:37–49

The Asymmetrical Evolution of the Mesopolyploid *Brassica oleracea* Genome

6

Chaobo Tong, Chittaranjan Kole, Lijiang Liu, Xiaohui Cheng, Junyan Huang, and Shengyi Liu

Abstract

The completion of *B. oleracea* genome sequencing has greatly enhanced *Brassica* research, especially for revealing genome structure and evolution of polyploidy. Based on intraspecies syntenic analysis and comparison with *B. rapa*, the asymmetrical evolution of polyploid *Brassica* genomes were elucidated at the DNA, RNA, and epigenetic levels. Main features include asymmetrical gene loss in chromosomal blocks derived from whole-genome triplication (WGT), asymmetrical sequence, and expression divergence among WGT-derived duplicated genes, asymmetrical 24-nt sRNA regulated subgenome expression dominance, asymmetrical TE (transposable elements) accumulation, gene expansion between two *Brassica* genomes, and asymmetrical contribution to the novelty of agronomic traits between the different subgenomes. It was found that the multilayered asymmetrical evolution is also observed in ancient polyploids such as Arabidopsis, maize, grape, soybean, grasses as well as recent allopolyploids such as rapeseed, cotton, and wheat. These evidences support the role of asymmetrical evolution in the evolutionary success of polyploidy as well as their phenotypic novelty and adaptation. Further understanding of the mechanism of asymmetrical evolution of polyploid genomes lays a foundation for genetic breeding and enhancement of elite cultivars.

C. Tong · L. Liu · X. Cheng · J. Huang · S. Liu (✉)
Key Laboratory of Biology and Genetic Improvement of Oil Crops, The Ministry of Agriculture and Rural Affairs, Oil Crops Research Institute, Chinese Academy of Agricultural Sciences, Wuhan 430062, China
e-mail: liusy@oilcrops.cn

C. Kole
Raja Ramanna Fellow, Department of Atomic Energy, Government of India, ICAR-National Institute for Plant Biotechnology, New Delhi, India

6.1 Duplicated Genomic Segments Derived from Whole-Genome Duplication in the *Brassica oleracea* Genome

Together with the release of the *Brassica rapa* genome sequence (Wang et al. 2011), the completion of the *B. oleracea* genome sequencing greatly enhanced the *Brassica* research, especially for revealing the genome structure and evolution of *Brassica* polyploid genome (Liu et al. 2014; Parkin et al. 2014). It was well documented that *Brassica* diploid species have been experienced a whole-genome triplication event in ~13 million years ago (Mya) (Liu et al. 2014; Lysak et al. 2005; Parkin et al. 2014; Wang et al. 2011). In addition, three old whole-genome duplication (WGD) events, the α, β, and

S. Liu et al. (eds.), *The Brassica oleracea Genome*, Compendium of Plant Genomes,
https://doi.org/10.1007/978-3-030-31005-9_6

γ events, also leave their legacy in *B. oleracea* genome (Leitch and Leitch 2008; Tang et al. 2008). Therefore, the *B. oleracea* contained the duplicated genomic segments derived from multiple rounds of WGDs, which was a good model for studying the polyploid genome evolution. Based on the syntenic analysis between *B. oleracea* and other genomes, the duplicated genomic segments contributed by WGD have been fully elucidated (Fig. 6.1). The distribution of genetic divergence among all duplicated genes exhibited multiple peaks corresponding to the WGD events of different ages. Although great gene losses and ongoing chromosomal rearrangement in the long-time evolution after these WGD events, the fractionated duplicated blocks and genes are still detected in *B. oleracea* genome. Through comparison with *A. thaliana* the triplicated genomic segments in *B. oleracea* corresponding to 24 ancestral blocks along *A. thaliana* chromosomes was identified (Schranz et al. 2006). Approximately 57.8% of *B. oleracea* genomic regions and 26,485 genes were included in these syntenic blocks. The syntenic analysis including identified duplicated chromosomal blocks and duplicated genes provided a foundation for studying the evolution of polyploid.

6.2 Asymmetrical Gene Loss After WGT in the *B. oleracea* Genome

A substantial loss of genes during the diploidization process following hexaploidy was observed in the *A. thaliana-B. oleracea* syntenic regions. If considering the 19,628 genes in *A. thaliana* as an initial gene count, WGT would result in the 58,884 (19,628 × 3) genes in three corresponding syntenic regions in *B. oleracea* genomes. However, only 26,485 genes were retained in the triplicated segments, showing a substantial gene loss. About 1.2 fold of ancestral genes after triplication were retained in the corresponding triplicated syntenic genomic regions. Moreover, the relative rates of gene loss/retention were asymmetrical among three subgenomes: 49.5% of the ancestral location have corresponding homologs retained in the least fractionated (LF) subgenome, whereas only 35.3% and 29.8% gene copies were maintained in moderately fractionated subgenome (MF1) and most fractionated subgenome (MF2), respectively (Fig. 6.2a). In the syntenic regions, about 54%, 36%, and 11% of ancestor genes retained to be one copy, two copies, and triplicates, respectively.

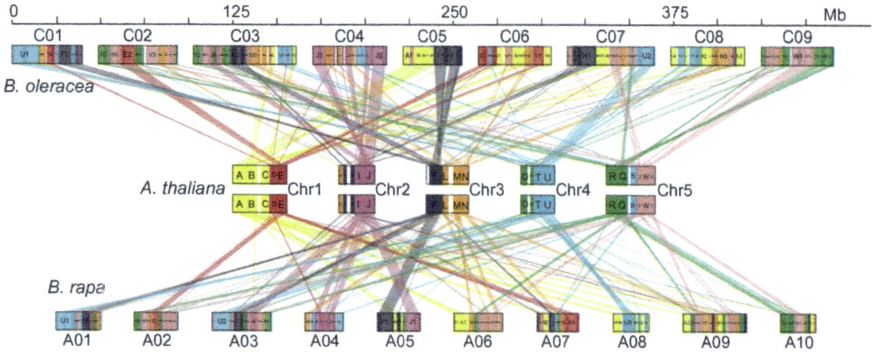

Fig. 6.1 The syntenic blocks between *A. thaliana*, *B. rapa* and *B. oleracea* (adopted from Liu et al. 2014). The A-X blocks were colored according to the previous definition in *A. thaliana*

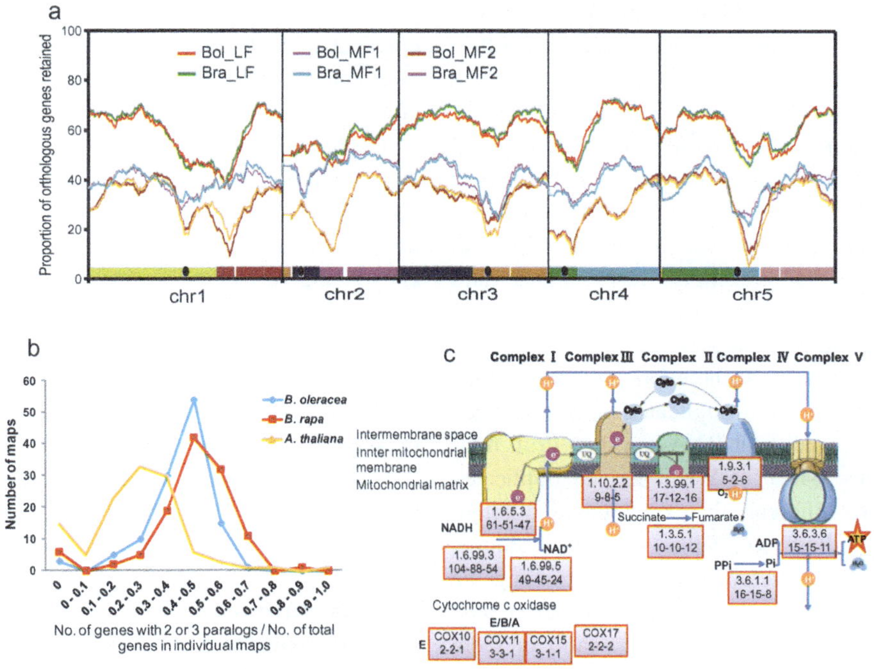

Fig. 6.2 Gene retention and loss in WGT-derived syntenic blocks of *B. oleracea* and *B. rapa* (adopted from Liu et al. 2014). **a** The frequency of ancestral gene retention/loss corresponding to *A. thaliana* A-X blocks in LF, MF1, and MF2 subgenomes. **b** The number of pathway maps contained WGT-derived duplicated genes in *A. thaliana*, *B. oleracea*, and *B. rapa*. For each map, the ratio of the number of WGT-derived duplicated genes was calculated. **c** WGT-derived duplicated genes of *Brassica* genomes in oxidative phosphorylation pathway. The gene number for each KO enzyme in *B. oleracea*, *B. rapa,* and *A. thaliana* were shown (dash-connected)

It is interesting to examine whether the retained genes preferentially existed in some special functional categories in *B. oleracea*. It was showed that WGT-produced paralogous genes were enriched in GO terms associated with regulation of transcription factors, RNA metabolism, and metabolic and biosynthetic processes. It also found that they are functionally enriched in pathways associated with energy metabolism such as carbon fixation, photosynthesis, oxidative phosphorylation, lipid metabolism, and circadian rhythm (Fig. 6.2b, c). In addition, the retained duplicated genes also showed over-retention in some gene families such as morphology specification (*TCP*), auxin functioning (*GH3, TIR, AUX, TAA, PIN, SAUR, IAA, TPL,* and *YUCCA*), and flowering time control (*FLC, AP1, LFY, GI, VRN1,* and *CO*). It is always observed that one Arabidopsis gene corresponds to two or three homologs in *B. rapa* and *B. oleracea* in these gene families. All these observations supported that the retained duplicated genes played an important role in the evolutionary innovation of polyploid and were associated with the morphological diversity of *Brassica* species.

6.3 Asymmetrical Sequence and Expression Divergence of WGT-Derived Duplicated Genes in the *B. oleracea* Genome

The divergence and differentiation of duplicated genes provided a foundation for the formation and evolution of polyploid. Their divergences in DNA and RNA levels were investigated, including gene structure, gene conversion, gene expression, and alternative splicing. At the DNA level, 38% (4,302/11,493) of WGT-derived duplicated genes in *B. oleracea* showed different exon numbers, exhibiting great sequence differentiation. Among these duplicated genes, some of them have suffered from nonreciprocal DNA exchanges (gene conversion). It showed that about 8% of duplicated quartets were generated from gene conversion in *B. oleracea*. Interestingly, LF subgenome has a higher ratio (about 44%) of gene conversion events, rather than MF1 and MF2 subgenomes.

At the RNA level, the expression divergence among WGT-derived duplicated genes has been revealed by analyzing RNA-seq data from multiple tissues. >40% of WGT-derived duplicated genes were differentially expressed in different tissues including root, callus, stem, leaf, silique, and flower of *B. oleracea*. By investigating the duplicated genes with high differentiated expression in GO categories, the ratio of them was significantly higher in GO terms related to catalytic activity, membrane, and defense response. In contrast, duplicated a less (38%) differentiated expression occurred in transcription factor gene pairs. Interestingly, genes in LF subgenomes always showed a high expression level when compared to the corresponding paralogs in MF1 and MF2 subgenomes (Fig. 6.3a). The relationship between expression dominance and high gene retention in LF subgenome could be explained by the following observation that there are less DNA methylation levels and other epigenetic regulations in LF subgenome (Cheng et al. 2016; Parkin et al. 2014; Woodhouse et al. 2014).

Furthermore, many duplicated genes showed expression differentiation by generating different transcripts. For the duplicated genes with same number of exons, >20% of them showed different splicing variants or the same splicing variants occurred differential expression in *B. oleracea* (Fig. 6.3b). As the different splicing variants may have different functions, the divergence of duplicated genes in AS variants provided a novel manner of polyploid evolution (Filichkin et al. 2010; Gabut et al. 2011; Syed et al. 2012; Zhang et al. 2010).

6.4 Asymmetrical Epigenetic Regulation of WGT-Derived Duplicated Genes in the *B. oleracea* Genome

Since DNA methylation is an important epigenetic modification regulating gene expression in *Brassica* species (Parkin et al. 2014), the DNA methylation level for all triplicated gene copies has been examined. It showed that the gene methylation levels were different among the three subgenomes, associated with subgenome expression dominance. It was found that 26% of

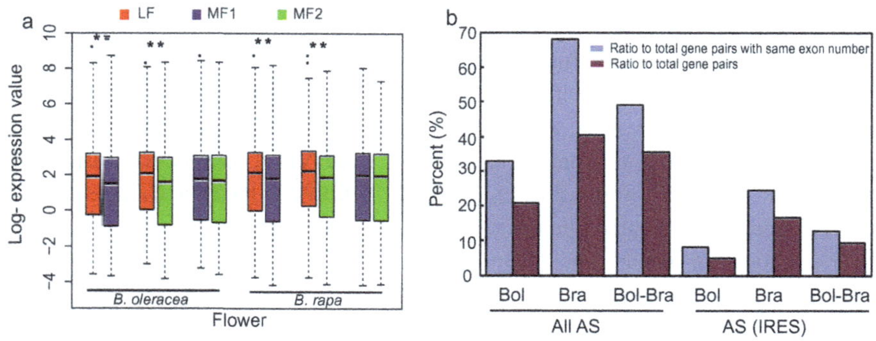

Fig. 6.3 The expression divergence of WGT-derived duplicated genes in *B. oleracea* and *B. rapa* (adopted from Liu et al. 2014). **a** The distribution of gene expression value of WGT-derived duplicated genes from LF, MF1, and MF2 subgenomes were showed in box and whisker plots. **b** The divergence of WGT-derived duplicated genes and Bol-Bra orthologous gene pairs on alternative splicing (AS) variants. The ratio of gene pairs with divergence of alternative splicing variants was calculated. The IRES represents the AS type of intron retention and exon shipping

all retained triplets showed different methylation levels among duplicated genes. Among all retained triplets, 43% of triplets showed high gene expression and low DNA methylation in LF subgenome, whereas 29% and 28% of triplets exhibited contrasted patterns in MF1 and MF2 subgenomes, respectively. This result showed that the methylation repatterning post-hybridization has been found to be asymmetrically distributed among the three subgenomes, which contributed to the establishment of subgenome expression dominance in the early evolutionary period after polyploidization. The gene expression could be suppressed when the 24-nt sRNA targeting TE existed in their promoter sequences (Hollister and Gaut 2009; Matzke and Mosher 2014; Zhang et al. 2014). A careful analysis of the relationships of transposable elements (TEs), mRNAs, and small RNAs among the three subgenomes have been conducted (Cheng et al. 2016; Woodhouse et al. 2014). It was revealed that the pattern of 24-nt sRNA induced DNA methylation was negatively significantly related to the gene expression among LF, MF1, and MF2 subgenomes. Compared to MF subgenomes, the LF subgenome with dominant expression has less distribution of 24-nt small RNAs target TEs. It was suggested that the biased distribution of 24-nt small RNAs among subgenomes was responsible for the formation of the subgenome dominance in *Brassica* species.

6.5 Asymmetrical TE Accumulation and Gene Expansion Between *B. oleracea* and *B. rapa*

A similar pattern of gene loss and retention after WGT was shared between *B. oleracea* and *B. rapa* because the majority of gene loss have occurred before the *B. oleracea*-*B. rapa* divergence. However, these two *Brassica* genomes experienced parallel evolution in the recent 4 million years, showing asymmetrical patterns in TE accumulation and gene expansion.

6.5.1 Asymmetrical TE Accumulation

It showed that TEs were asymmetrically amplified between *B. oleracea* and *B. rapa* genomes. The retro- and DNA TEs occupied 22.13% and 16.67% of *B. oleracea* genome, respectively, significantly higher than that in *B. rapa* genome (9.43 and 12.04%). Due to more amplified TEs in *B. oleracea*, the length of *B. oleracea*-*B. rapa* syntenic region in *B. oleracea* is 3.4 times longer than counterparts in *B. rapa*. The asymmetrical amplification between *B. oleracea* and *B. rapa* was also confirmed by phylogenetic analysis of long terminal repeats (LTR-RTs). It showed that more LTR-RTs clustered in most families in *B. oleracea* than *B. rapa*. The insertion time estimation of LTR-RTs showed that majority of intact LTR-RTs amplified continuously in *B. oleracea* genome after divergence from *B. rapa*. Whereas LTR-RTs only actively amplified in the last 1 million years, especially in the recent 0.2 MY (Fig. 6.4a). Due to the asymmetrical accumulation between *B. oleracea* than *B. rapa*, more LTR-RTs are distributed in euchromatic regions of *B. oleracea* than *B. rapa* (Fig. 6.4b).

6.5.2 Asymmetrical Gene Expansion

Although *B. oleracea* and *B. rapa* are highly similar in gene order and number in syntenic regions, a large number of species-specific genes were also identified in two *Brassica* genomes. There were 9,832 and 5,735 species-specific genes in *B. oleracea* and *B. rapa* (Fig. 6.5a). Ninety percent of species-specific genes were validated for their absence by the alignment of Illumina raw reads. It was found that species-specific genes were dispersed and surrounded by non-specific genes along chromosomes, suggesting that the individual gene loss may be the main reason leading to the formation of species-specific genes. Additionally, tandem duplication also produces many species-specific genes in *Brassica* genomes. There are 1,554, 1,825, and

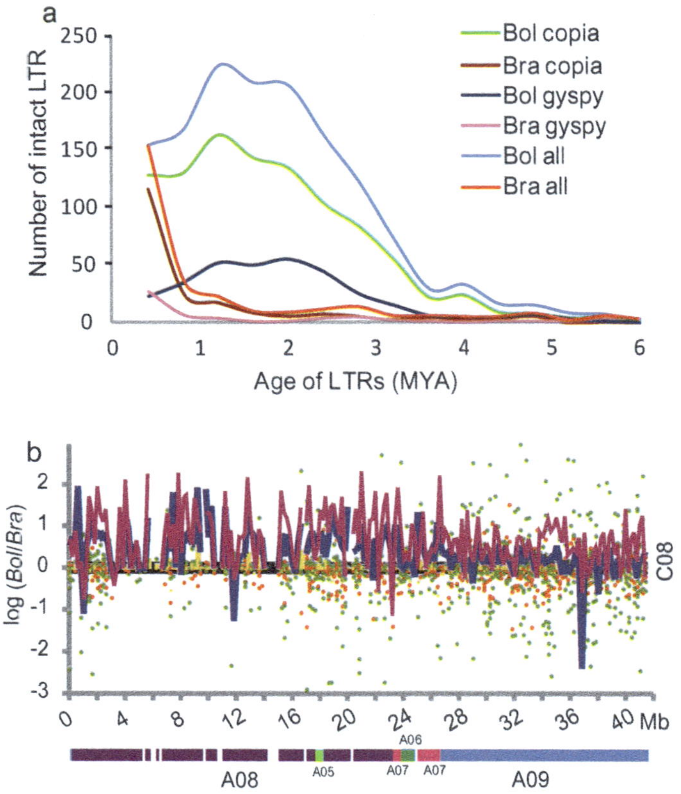

Fig. 6.4 The asymmetrical TE amplification between *B. oleracea* and *B. rapa* (adopted from Liu et al. 2014). **a** The distribution of insertion time of LTR-RTs (Copia and Gypsy) in syntenic regions between *B. rapa* and *B. oleracea*. **b** The number and composition compared in syntenic blocks between *B. oleracea* and *B. rapa*. The ratio for intron length, exons length, gene length, LTR length, and total syntenic block length was calculated by Log10 (*B. oleracea*/*B. rapa*), denoted as green point, red point, yellow point, purple line, and blue line, respectively

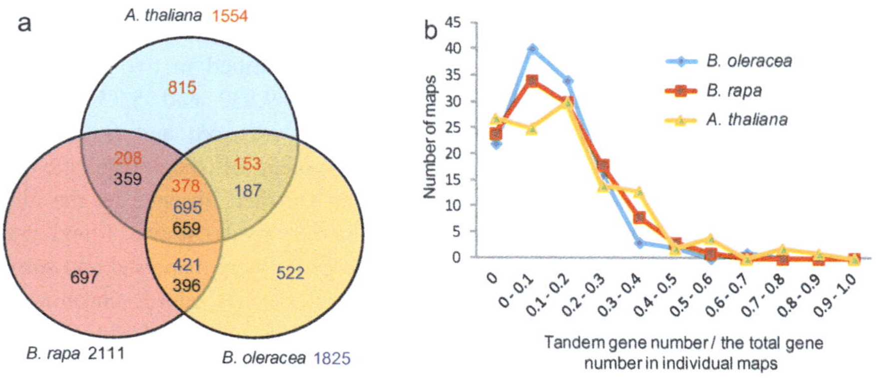

Fig. 6.5 The duplicated genes derived from tandem duplication and WGD in *Brassica* genomes (adopted from Liu et al. 2014). **a** The shared and specific genes derived from tandem duplication was shown in *A. thaliana*, *B. rapa*, and *B. oleracea*. **b** The number of KEGG maps with different ratios of tandem gene number relative to total gene number in individual maps

2,111 gene clusters in *A. thaliana*, *B. oleracea* and *B. rapa* and which contained 4,170, 4,365, and 5,181 tandemly duplicated genes, respectively. Among them, 815, 522, and 697 tandem clusters were detected as species-specific in *A. thaliana*, *B. oleracea,* and *B. rapa* genomes. These species-specific tandemly duplicated genes were enriched in GO terms associated with defense response and secondary metabolism such as tropane, indole alkaloid biosynthesis, pyridine alkaloid biosynthesis, and piperidine (Fig. 6.5b).

6.6 Asymmetrical Evolution of Polyploid Plant Genomes

Majority of flowering plants contained the legacy of polyploidization as many old WGD events widely occurred during plant evolution (Amborella Genome 2013). After polyploidization, the genome structure and function will be changed to initiate the diploidization process (Chen 2007) including gene loss and the divergence in sequence and expression of retained duplicated genes (Adams and Wendel 2005). During the diploidization processes, the different subgenomes may be subjected to the asymmetrical evolution including DNA, expression, and epigenetic levels, as is the case in WGT-derived *Brassica* genomes. The asymmetrical evolution of subgenomes in polyploid provided a basis for the formation of polyploid morphological and biological complexity (Soltis et al. 2009).

In *Brassica* species, the asymmetrical gene loss was observed among three subgenomes (LF, MF1, and MF2). The subgenome dominance has been found among them where the ancestral genes after duplication were preferentially retained in LF subgenome, rather than MF1 and MF2 subgenomes. The phenomenon of subgenome dominance has also been reported in Arabidopsis (Thomas et al. 2006), maize (Schnable et al. 2012b), grasses (Schnable et al. 2012a). Furthermore, this type of asymmetrical gene loss was also found in young allopolyploids such as rapeseed, cotton, and wheat, indicating that it can occur in a much shorter evolutionary time. In rapeseed, the frequency of gene losses in A subgenome was obviously higher than the C subgenome (Chalhoub 2014). In cotton, the rate of gene loss in the D subgenome was obviously faster than A subgenome (Fang et al. 2017; Zhang et al. 2015). In wheat, B subgenome owned most gene loss and arrangement which was considered as most plastic genomes (El Baidouri et al. 2017; Pont et al. 2013; Pont and Salse 2017), while the young D subgenome was considered as the most stable subgenome (Zhang et al. 2013).

The asymmetrical accumulation of TEs among different subgenomes was also observed in many polyploid genomes, however, the difference seems to be derived from the asymmetrical proliferation of TEs between closely related diploids. In *B. napus*, TEs were asymmetrically distributed in A_n and C_n subgenomes, (Chalhoub 2014), consistent with the greater extent of TE amplification in *B. oleracea* rather than *B. rapa* (Liu et al. 2014; Parkin et al. 2014). In allotetraploid cotton, the A subgenome was obviously larger than D subgenome because the TEs were more actively amplified and expanded in A subgenome (Zhang et al. 2015). In wheat, the TEs have amplified with a high level in B subgenome and resulted in a large number of structural variations (El Baidouri et al. 2017; Pont et al. 2013). These observations suggested that the asymmetrical amplification of TEs between different subgenomes in polyploid genomes became an important manner during the diploidization processes (Casacuberta and Gonzalez 2013).

Significant asymmetrical expression dominance among WGT-derived subgenomes was observed in *B. oleracea,* in agreement with findings in paleopolyploid maize (Schnable et al. 2011) and allotetraploid Arabidopsis (Wang et al. 2006). However, the differences in gene expression have not been reported at the whole subgenomes level in the recent polyploid genomes such as rapeseed, cotton, and wheat. In *B. napus*, although about 40% of duplicated genes were diverged between A_n and C_n subgenomes but there is no expression bias at the whole subgenome level (Chalhoub 2014). In cotton, the subgenomic dominance in expression was also not observed between A and D subgenomes although 20–40% of duplicated genes showed biased expression in A or D subgenome (Adams et al. 2003; Flagel and Wendel 2010). In wheat, there are 27% of triplets between A, B, and D homoeologs diverged in expression when considering a single tissue, and this ratio is up to 81% when considering many tissues together (Bottley et al. 2006; Mochida et al. 2006).

The epigenetic modification was considered as an important mechanism for regulating the asymmetrical divergences of duplicated genes in polyploid genomes (Hollister and Gaut 2009; Rodin and Riggs 2003). In *Brassica* species, the small RNA-targeted TE methylation was reported to be responsible for the subgenomic dominance and biased gene loss among different subgenomes (Cheng et al. 2016; Woodhouse et al. 2014). From these studies, it was suggested that TE may drive subgenome structural differentiation and play an important role in stabilizing meiotic pairing and diploidization processes. The asymmetrical TE accumulation among different progenitors and inheritance through polyploidy may drive gene silencing and following gene loss after polyploidization. This asymmetrical evolution after polyploidization created genomic plasticity and also ultimately resulted in the novelty of phenotypes or agronomic traits caused by polyploidy. In *B. napus*, subgenome A contributed more genomic regions significantly associated with traits including yield, oil content, flowering time, while subgenome C mainly controls the responses to biotic and abiotic factors (Qian et al. 2014). The different subgenomes of cotton also were subjected to asymmetrical selection and domestication, showing that subgenome A owns more significantly positively selected genes (PSGs) (Fang et al. 2017; Zhang et al. 2015). It was found that the functional genes controlling fiber development were preferentially located in the A-genome progenitor, but majority of the genes related to plant adaptation distributed in the D subgenome (Fang et al. 2017; Wang et al. 2017). In wheat, genes related to morphological traits were found to be concentrated on A subgenome, while genes associated with biotic and abiotic stress mainly distributed in B subgenome (Feldman and Levy 2012). These results supported that symmetrical evolution provided a solid genetic basis for the polyploidy evolution as well as their phenotypic novelty and adaptation. Further understanding the mechanism of asymmetrical evolution of polyploid genomes will provide a foundation for genetic breeding and enhancement of elite cultivars.

References

Adams KL, Cronn R, Percifield R, Wendel JF (2003) Genes duplicated by polyploidy show unequal contributions to the transcriptome and organ-specific reciprocal silencing. Pro Natl Acad Sci USA 100:4649–4654

Adams KL, Wendel JF (2005) Polyploidy and genome evolution in plants. Curr Opin Plant Biol 8:135–141

Amborella Genome P (2013) The Amborella genome and the evolution of flowering plants. Science 342:1241089

Bottley A, Xia GM, Koebner RM (2006) Homoeologous gene silencing in hexaploid wheat. Plant J 47:897–906

Casacuberta E, Gonzalez J (2013) The impact of transposable elements in environmental adaptation. Mol Ecolo 22:1503–1517

Chalhoub B (2014) Early allopolyploid evolution in the post-Neolithic Brassica napus oilseed genome (vol 348, 1260782, 2014). Science 345:1255–1255

Chen ZJ (2007) Genetic and epigenetic mechanisms for gene expression and phenotypic variation in plant polyploids. Annu Rev Plant Biol 58:377–406

Cheng F, Sun C, Wu J, Schnable J, Woodhouse MR, Liang J, Cai C, Freeling M, Wang X (2016) Epigenetic regulation of subgenome dominance following whole genome triplication in *Brassica rapa*. New Phytol 211:288–299

El Baidouri M, Murat F, Veyssiere M, Molinier M, Flores R, Burlot L, Alaux M, Quesneville H, Pont C, Salse J (2017) Reconciling the evolutionary origin of bread wheat (*Triticum aestivum*). New Phytol 213:1477–1486

Fang L, Guan XY, Zhang TZ (2017) Asymmetric evolution and domestication in allotetraploid cotton (*Gossypium hirsutum* L.). Crop J 5:159–165

Feldman M, Levy AA (2012) Genome evolution due to allopolyploidization in wheat. Genetics 192:763–774

Filichkin SA, Priest HD, Givan SA, Shen R, Bryant DW, Fox SE, Wong WK, Mockler TC (2010) Genome-wide mapping of alternative splicing in *Arabidopsis thaliana*. Genome Res 20:45–58

Flagel LE, Wendel JF (2010) Evolutionary rate variation, genomic dominance and duplicated gene expression evolution during allotetraploid cotton speciation. New Phytol 186:184–193

Gabut M, Samavarchi-Tehrani P, Wang X, Slobodeniuc V, O'Hanlon D, Sung HK, Alvarez M, Talukder S, Pan Q, Mazzoni EO et al (2011) An alternative splicing switch regulates embryonic stem cell pluripotency and reprogramming. Cell 147:132–146

Hollister JD, Gaut BS (2009) Epigenetic silencing of transposable elements: a trade-off between reduced transposition and deleterious effects on neighboring gene expression. Genome Res 19:1419–1428

Leitch AR, Leitch IJ (2008) Genomic plasticity and the diversity of polyploid plants. Science 320:481–483

Liu S, Liu Y, Yang X, Tong C, Edwards D, Parkin IA, Zhao M, Ma J, Yu J, Huang S et al (2014) The *Brassica oleracea* genome reveals the asymmetrical evolution of polyploid genomes. Nat Commun 5:3930

Lysak MA, Koch MA, Pecinka A, Schubert I (2005) Chromosome triplication found across the tribe Brassiceae. Genome Res 15:516–525

Matzke MA, Mosher RA (2014) RNA-directed DNA methylation: an epigenetic pathway of increasing complexity. Nat Rev Genet 15:394–408

Mochida K, Kawaura K, Shimosaka E, Kawakami N, Shin IT, Kohara Y, Yamazaki Y, Ogihara Y (2006) Tissue expression map of a large number of expressed sequence tags and its application to in silico screening of stress response genes in common wheat. Mol Genet Genomics MGG 276:304–312

Parkin IA, Koh C, Tang H, Robinson SJ, Kagale S, Clarke WE, Town CD, Nixon J, Krishnakumar V, Bidwell SL et al (2014) Transcriptome and methylome profiling reveals relics of genome dominance in the mesopolyploid *Brassica oleracea*. Genome Biol 15:R77

Pont C, Murat F, Guizard S, Flores R, Foucrier S, Bidet Y, Quraishi UM, Alaux M, Dolezel J, Fahima T et al (2013) Wheat syntenome unveils new evidences of contrasted evolutionary plasticity between paleo- and neoduplicated subgenomes. Plant J 76:1030–1044

Pont C, Salse J (2017) Wheat paleohistory created asymmetrical genomic evolution. Curr Opin Plant Biol 36:29–37

Qian LW, Qian W, and Snowdon RJ (2014) Sub-genomic selection patterns as a signature of breeding in the allopolyploid *Brassica napus* genome. Bmc Genomics 15

Rodin SN, Riggs AD (2003) Epigenetic silencing may aid evolution by gene duplication. J Mol Evol 56:718–729

Schnable JC, Freeling M, Lyons E (2012a) Genome-wide analysis of syntenic gene deletion in the gasses. Genome Biol Evol 4:265–277

Schnable JC, Springer NM, Freeling M (2011) Differentiation of the maize subgenomes by genome dominance and both ancient and ongoing gene loss. Proc Natl Acad Sci U S A 108:4069–4074

Schnable JC, Wang X, Pires JC, Freeling M (2012b) Escape from preferential retention following repeated whole genome duplications in plants. Front Plant Sci 3:94

Schranz ME, Lysak MA, Mitchell-Olds T (2006) The ABC's of comparative genomics in the Brassicaceae: building blocks of crucifer genomes. Trends Plant Sci 11:535–542

Soltis DE, Albert VA, Leebens-Mack J, Bell CD, Paterson AH, Zheng CF, Sankoff D, dePamphilis CW, Wall PK, Soltis PS (2009) Polyploidy and angiosperm diversification. Am J Bot 96:336–348

Syed NH, Kalyna M, Marquez Y, Barta A, Brown JW (2012) Alternative splicing in plants–coming of age. Trends Plant Sci 17:616–623

Tang H, Wang X, Bowers JE, Ming R, Alam M, Paterson AH (2008) Unraveling ancient hexaploidy through multiply-aligned angiosperm gene maps. Genome Res 18:1944–1954

Thomas BC, Pedersen B, Freeling M (2006) Following tetraploidy in an Arabidopsis ancestor, genes were removed preferentially from one homeolog leaving clusters enriched in dose-sensitive genes. Genome Res 16:934–946

Wang J, Tian L, Lee HS, Wei NE, Jiang H, Watson B, Madlung A, Osborn TC, Doerge RW, Comai L et al (2006) Genome-wide non-additive gene regulation in Arabidopsis allotetraploids. Genetics 172:507–517

Wang MJ, Tu LL, Lin M, Lin ZX, Wang PC, Yang QY, Ye ZX, Shen C, Li JY, Zhang L, et al (2017) Asymmetric subgenome selection and cis-regulatory divergence during cotton domestication. Nat Genetics 49:579

Wang X, Wang H, Wang J, Sun R, Wu J, Liu S, Bai Y, Mun JH, Bancroft I, Cheng F et al (2011) The genome of the mesopolyploid crop species Brassica rapa. Nat Genet 43:1035–1039

Woodhouse MR, Cheng F, Pires JC, Lisch D, Freeling M, Wang XW (2014) Origin, inheritance, and gene regulatory consequences of genome dominance in polyploids. Proc Natl Acad Sci USA 111:5283–5288

Zhang H, Zhu B, Qi B, Gou X, Dong Y, Xu C, Zhang B, Huang W, Liu C, Wang X et al (2014) Evolution of the BBAA component of bread wheat during its history at the allohexaploid level. Plant cell 26:2761–2776

Zhang HK, Bian Y, Gou XW, Dong YZ, Rustgi S, Zhang BJ, Xu CM, Li N, Qi B, Han FP et al (2013) Intrinsic karyotype stability and gene copy number variations may have laid the foundation for tetraploid wheat formation. Proc Natl Acad Sci USA 110:19466–19471

Zhang PG, Huang SZ, Pin AL, Adams KL (2010) Extensive divergence in alternative splicing patterns after gene and genome duplication during the evolutionary history of Arabidopsis. Mol Biol Evol 27:1686–1697

Zhang TZ, Hu Y, Jiang WK, Fang L, Guan XY, Chen JD, Zhang JB, Saski CA, Scheffler BE, Stelly DM et al (2015) Sequencing of allotetraploid cotton (*Gossypium hirsutum* L. acc. TM-1) provides a resource for fiber improvement. Nat Biotechnol 33:531–U252

Asymmetric Evolution of Transposable Elements in *Brassica oleracea*

7

Meixia Zhao, Shengyi Liu, and Jianxin Ma

Abstract

The reference sequences of the two *Brassica* diploids, *Brassica rapa* and *Brassica oleracea*, and their allotetroploid, *Brassica napus,* demonstrate that transposable elements (TEs) are the major constituent of these three *Brassica* species and function as the main contributor to the genome size variation. Although TEs have the ability to proliferate to high copies in the genome, most of them are suppressed by epigenetic pathways and are removed by DNA removal mechanisms, such as unequal homologous recombination and illegitimate recombination. Recent advances indicate the accumulation, elimination, and evolutionary rates of different types of TEs are dynamically distinct within the *B. oleracea* genome and among different *Brassica* species. The responses of TEs to polyploidy or interspecific hybridization remain largely unknown in *Brassica*. Nevertheless, TEs have been known to produce varieties of changes in gene expression and evolution, suggesting the importance of TEs in the regulation of *Brassica* species.

7.1 Introduction

Transposable elements (TEs) are the major DNA component in flowering plants, with the properties of self-replication and reposition, and contribute significantly to the genome size variation and structure differentiation even between closely related species. The two major classes of TEs are those that transpose via an RNA intermediate in a 'copy-and-paste' process, designated as retrotransposons or class I elements, and those that employ a 'cut-and-paste' or replicate via a rolling-circle mechanism, named DNA transposons or class II elements (Finnegan 1989; Wicker et al. 2007). Retrotransposons are further grouped into long terminal repeat retrotransposons (LTR-RTs), including two main superfamilies, *Ty1/Copia* and *Ty3/Gypsy*, and non-LTR retrotransposons, consisting of long interspersed nucleotide elements (LINEs) and short interspersed nucleotide elements (SINEs) (Grandbastien 1992; Kumar and Bennetzen 1999; Lander et al. 2001). Unlike retrotransposons, DNA transposons, such as *Tc1/Mariner*,

M. Zhao (✉)
Department of Biology, Miami University, 700 E. High Street, Oxford, OH 45056, USA
e-mail: meixiazhao@miamioh.edu

S. Liu
Key Laboratory of Biology and Genetic Improvement of Oil Crops, The Ministry of Agriculture and Rural Affairs, Oil Crops Research Institute, Chinese Academy of Agricultural Sciences, Wuhan 430062, China

J. Ma
Department of Agronomy, Purdue University, 915 W. State Street, West Lafayette, IN 47907, USA

S. Liu et al. (eds.), *The Brassica oleracea Genome*, Compendium of Plant Genomes,
https://doi.org/10.1007/978-3-030-31005-9_7

hAT, CACTA, *Mutator*-like, *PIF/Harbinger*, and *Pong* elements transpose via the excision/repair of one site and the integration of a new site either nearby on the same chromosome or elsewhere in the genome (Wicker et al. 2007; Fedoroff 2012). However, a few groups of DNA transposons, such as *Helitron* and *Maverick* elements do not completely obey the 'cut-and-paste' replication and the transposition model. It was found that *Helitron* elements follow a rolling-circle transposition strategy, and only one strand is transferred from one genomic site to another, mimicking DNA-virus and bacteria replication mechanisms (Kapitonov and Jurka 2001, 2006; Hollister and Gaut 2007; Pritham and Feschotte 2007).

In addition to polyploidization, rapid proliferation of TEs, particularly LTR-RTs, is the primary mechanism driving plant genome expansion (Bennetzen et al. 2005). For instance, the large 2.3 Gb maize genome is comprised of 85% repetitive DNA, of which 75% are LTR-RTs with several high copy number families (Schnable et al. 2009; Jiao et al. 2017). While the small 115 Mb *Aarabidopsis thaliana* genome might be lacking in transposon proliferation, with ~10% repeats (The Arabidopsis Genome Initiative 2000). Given this, TEs were first called 'junk DNA' or 'parasites'. However, further understanding of how TEs shape genomic structures, how TEs affect gene expression and genome evolution, and their epigenetic regulation indicates that TEs are far from being considered as 'junk' or 'selfish' DNA sequences, and thus were called 'controlling elements' by Barbara McClintock (McClintock 1956a, b). TEs are evidently proved to be powerful sources of genetic changes and have played an important role in plant genome evolution.

Brassica oleracea (CC, $2n = 18$) belongs to the Brassicaceae family, which consists of 338 genera and approximately 3,700 species in 25 tribes (Beilstein et al. 2006). Within recent years, five out of the six species in the famous 'triangle of U', including the three *Brassica* diploids (*B. oleracea*, *Brassica rapa*, AA, $2n = 20$, and *Brassica nigra*, BB, $2n = 16$) and their allotetraploid *Brassica napus* (AACC, $2n = 38$) and *Brassica juncea* (AABB, $2n = 36$), have been continuously sequenced (Wang et al. 2011a; Liu et al. 2014; Chalhoub et al. 2014; Yang et al. 2016). The availability of the genomic data has provided unprecedented opportunities for detailed comparisons of TEs among different *Brassica* species. Genome size evaluation of *B. rapa* (~529 Mb) and *B. oleracea* (~696 Mb) showed that the size variation of these two *Brassica* genomes is about ~170 Mb (Johnston et al. 2005). The relative expansion was probably caused by the differential accumulation of TEs in the orthologous regions investigated (Cheung et al. 2009). In this chapter, with the comparison to *B. rapa*, we will discuss the abundance and distribution of TEs, the dynamics of TE proliferation and elimination, the molecular mechanisms involved in the asymmetric evolution of TEs, and the potential effects of TEs on gene expression in the *B. oleracea* genome.

7.2 TE Abundance

It was suggested that *A. thaliana* and *B. oleracea* diverged from a common ancestor approximately 20 million years ago (MYA) (Yang et al. 1999). A whole-genome triplication (WGT) event only occurred on the *Brassica* lineage after the split from *Arabidopsis* (Yang et al. 2006). However, this WGT event could not fully explain the five-fold genome size difference between *B. oleracea* and *A. thaliana*. Comparison of the whole genome sequence of *A. thaliana* with a partial draft of the *B. oleracea* genome reveals that all class I and a few families of class II elements, such as *Pong*-like and CACTA-like transposons, have amplified to a very high copy number in *B. oleracea* (Zhang and Wessler 2004). Recent whole-genome sequencing indicates that 38.8% of the ~540 Mb assembled genome of *B. oleracea* is composed of TE-related sequences, which contains 5,107 retrotransposons, 8,275 DNA transposons with complete structures, and tens of thousands of truncated TE fragments (Fig. 7.1; Table 7.1; Zhao et al. 2013; Liu et al. 2014). The significantly lower gene density and higher TE density were detected in *B. oleracea* than those in *B. rapa* (Qiu et al. 2009; Zhao et al. 2013). It was proposed that TEs were able to explain about 60% of the size difference of the syntenic regions between *B. oleracea* (88.3 Mb of 259.6 Mb) and *B. rapa* (25.9 Mb of 155.0 Mb) during the 3.7 million year evolution (Inaba and Nishio 2002; Liu et al. 2014).

7.3 Class I Retrotransposons

Retrotransposons are amplified via an RNA intermediate, which could rapidly increase copy numbers in the parasitized genome. Based on the currently assembled reference genome of *B. oleracea*, it was demonstrated that retrotransposons, including 4,614 LTR-RTs, 140 LINEs, and 353 SINEs, together with their truncated remnants, occupy 22.13% genomic DNA of the assembled sequences of *B. oleracea* (Fig. 7.1; Table 7.1; Zhao et al. 2013; Liu et al. 2014). Before the *B. oleracea* reference sequence was released, Alix and Heslop-Harrison (2004) analyzed the diversity of retrotransposons from 16 accessions of the six *Brassica* species in the triangle of U, and isolated 80 fragments of reverse transcriptase domains. Further investigation on five representative reverse transcriptase domains, including two *Ty1/Copia*-like, two *Ty3/Gypsy*-like (one associated with Athila lineage), and one LINE element in *B. oleracea*, reveals a large variation of the abundance and chromosome distribution of these five retroelement families. The estimated copy numbers of these four LTR-RT families are between 60 and 570; however, LINE is present at a very low copy in the *B. oleracea* genome. Fluorescent in situ hybridization indicates that the Athila-like *gypsy* element has preferential insertions into pericentromeric regions, suggesting that this element may be involved in centromere evolution (Alix et al. 2005). Plant centromere is generally composed of a large number of centromeric satellite arrays and centromere-specific retrotransposons (CRs), such as CentO and CRRs in rice, and CentC and CRMs in maize (Zhong et al. 2002; Jiang et al. 2003; Nagaki et al. 2004; Ma et al. 2007). There are two classes of centromeric satellite repeats named CentBr1 and CentrBr2 identified in *B. rapa* and *B. oleracea*. CentBr1 and CentrBr2 are with a length of 176 bp and share 86% sequence identity (Harrison and Heslop-Harrison 1995; Lim et al. 2005). Centromere-specific retrotransposon (CRB) and various pericentromere-

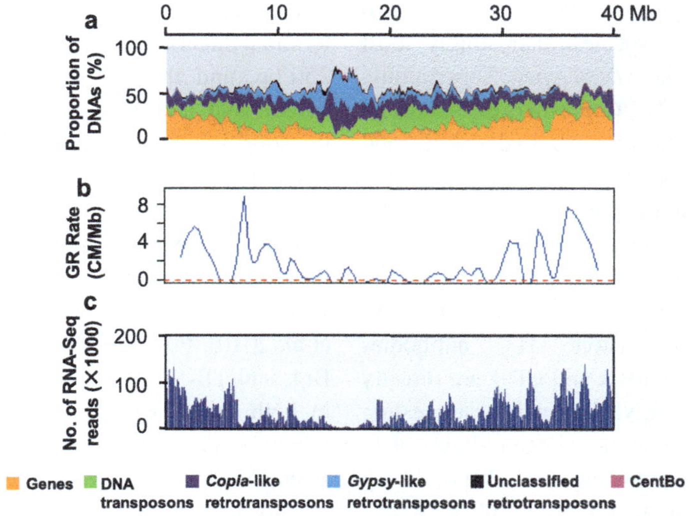

Fig. 7.1 Genomic features and expression values along chromosome 9 of *B. oleracea*. **a** Distribution of the proportions of transposable elements, genes and centromere satellite repeats Adapted from Liu et al. 2014. **b** Distribution of local recombination rates. The data used here was extracted from Wang et al. 2012. **c** Distribution of expression values of leaf tissue

Table 7.1 Transposable elements with clear boundaries and signatures of insertion sites identified in the assembled portions of the *Brassica rapa* and *Brassica oleracea* genomes

Classification			Copy numbers	
			B. rapa	*B. oleracea*
Class I: Retrotransposons			1,332	5,107
	LTR-RT		1,039	4,614
		LTR-RT/*Copia*	665	2,952
		LTR-RT/*Gypsy*	291	1,382
		LTR-RT/Unclassified	83	280
	Non-LTR-RT		293	493
		LINE	81	140
		SINE	212	353
Class II: DNA transposons			3,270	8,275
	Subclass I:		2,657	6,663
		Tc1/Mariner	75	76
		hAT	277	560
		Mutator	235	521
		PIF/Harbinger	55	69
		Pong	607	931
		CACTA	503	2,519
		MITE/Stowaway	636	1,474
		MITE/Tourist	269	513
	Subclass II:		613	1,612
		Helitron	613	1,612
Total			4,602	13,382

specific retrotransposons (PCRBr) were also identified in *Brassica* species. Interestingly, most CRs belong to the *Ty3/Gypsy* superfamily, whereas CRB is a *Ty1/Copia*-like element, and presents in all of the three diploid species and their allotetraploid relatives. PCRBr was only found in the A genomes/subgenomes, suggesting that it is A-genome-specific (Lim et al. 2007). Further Chromatin immunoprecipitation and immunolabeling tests with anti-BrCENH3 (centromere-specific histone H3) antibodies indicate that both CentBrs and CRB are directly associated with BrCENH3 proteins, suggesting that they are part of functional centromeres of *B. rapa* and *B. oleracea* (Wang et al. 2011b).

A group of non-autonomous LTR-RTs, termed terminal-repeat retrotransposons in miniature (TRIM) elements have been identified in *Brassica* relatives (Yang et al. 2007). These TRIM elements are typically short, with terminal direct repeat sequences between 100 and 250 bp, and the internal parts 100–300 bp in length, and lack essential proteins required for transposition and integration (Witte et al. 2001). Although TRIM elements are not able to autonomously amplify and transpose, they are thought to have transposition activities with the help of their autonomous partners (Jiang et al. 2002a, b; Kalendar et al. 2004; Kejnovsky et al. 2006; Du et al. 2010). Five TRIM families (Br1, Br2, Br3, Br4, and TB-5) were identified with high copies in both the *B. rapa* and *B. oleracea* genomes (Yang et al. 2007; Murukarthick et al. 2014). Interestingly, more than 40% of those elements are located within 2 kb regions of protein-encoding genes, indicating that they might have potential effects on gene expression (Kwon et al. 2007; Murukarthick et al. 2014). However, because their autonomous partner has not been found, the detailed mechanisms involved in how these TRIM elements utilize the proteins encoded by the autonomous elements and what the origins of these non-autonomous elements are remain unclear.

SINEs are short (80–500 bp) non-autonomous elements and normally reply on autonomous LINEs for retransposition. Nonetheless, they are not deletion derivatives of autonomous LINE retrotransposons (Wicker et al. 2007). Several SINE families are present in *Brassicaceae*, including S1(SB1), RathE1, RathE2, and RathE3 (Deragon et al. 1994; Myouga et al. 2001; Lenoir et al. 2005; Murukarthick et al. 2014). Evolutionary analysis of SINE families between *A. thaliana* and *B. oleracea* indicates that SINE elements were recently amplified in the *Brassica* lineages (Deragon and Zhang 2006). The first SINE family discovered in *Brassicaceae* is S1, which is approximately 180 bp long, and distributes widely in the whole tribe (Deragon et al. 1994; Lenoir et al. 1997; Tatout et al. 1999). S1 contains the typical structures of SINE family, such as 5' tRNA 'head', internal conserved RNA polymerase III (Pol III) promoter (including A and B boxes), and 3' Poly(A) 'tail' (Deragon et al. 1996). Further analysis of the Pol III promoter indicates that, although the S1 promoter can be functional indepdently, the external sequences can enhance the transcriptional level of the S1 elements (Arnaud et al. 2001). S1 SINE retrotransposons were highly methylated, not only in the transcriptionally important regions but also along the whole body of the elements. Moreover, the methylation has not resulted from the chromatin environment, suggesting that the S1 elements were especially methylated (Goubely et al. 1999). The insertion variability of S1 amplification was used as a classification criterion to evaluate phylogenetic relationships on closely related species from the *B. oleracea* species (Tatout et al. 1999). Thus, SINE retrotransposons tend to be very useful for researchers to understand the evolutionary history of *Brassica* species, and could also be used to develop molecular markers for gene isolation and plant breeding.

7.4 Class II DNA Transposons

DNA transposons are generally less abundant than LTR-RTs in plant genomes. In *B. oleracea*, approximately 16.67% of the genomic DNA is class II transposon-like sequences (Liu et al. 2014). Interestingly, CACTA-like elements make up the largest fraction of the DNA transposons in the *B. oleracea* genome. Of the 8,275 DNA transposons with complete structures, 2,519 are CACTA-like elements (Table 7.1; Zhao et al. 2013). Previous investigation identified a *Brassica* CACTA element *Bot1* (*Brassica oleracea transposon 1*) and found that it has extensively proliferated within the C genome. Moreover, it was indicated that this element contains a C genome-specific domain, and carries a host S locus-associated *SLL3* gene, which was also involved in the several rounds of *Bot1* amplification (Alix et al. 2008). Further, whole-genome analysis identified ~603 intact copies belonging to *Bot1* family in *B. oleracea* (named *Boc1*, *Brassica olearcea CACTA family 1*), while only ~123 copies were found in *B. rapa* (named *Brc1*, *Brassica rapa CACTA family 1*), suggesting that *Bot1* played an important role in the genome size differentiation and genome divergence of the A and C *Brassica* genomes (Zhao et al. 2013).

Miniature inverted-repeat transposable elements (MITEs) are a class of non-autonomous DNA transposons, generally with a length shorter than 1 kb. In plants, MITEs fall into two major superfamilies, *Tourist*-like, which originate from *PIF/Harbinger* or *Pong* elements, and *Stowaway*-like, which are derivates of *Tc1/Mariner* transposons (Bureau and Wessler 1992, 1994). Unlike other class II elements, MITEs can proliferate to tens of thousands of copies in the genome. For instance, around 90,000 MITEs are present in the rice genome. Distribution analysis of these MITE elements reveals that they preferentially insert into gene-rich chromosomal arm regions, suggesting that they may have potential effects on gene expression (Jiang et al. 2004a, b). MITEs were also found to be ubiquitous in *Brassica*. It was estimated that there are about

30,000 MITE-related sequences with large insertion polymorphisms in *B. rapa* and *B. oleracea* (Nouroz et al. 2015). A *Stowaway*-like MITE, *BraSto*, was first characterized in *B. rapa*. Further analysis indicates that *BraSto* elements were primarily located near or within genes, and were still active in the two *Brassica* diploids and their allotetraploid *B. napus* genome. Frequent associations of *BraSto* with putative promoters and regulatory regions of genes and the enrichment of *BraSto* sequences in binding sites for transcription factors may shed new light on their significance on gene expression (Sarilar et al. 2011). Another *Stowaway*-like MITE family *BRAMI*-1 was reported to be present in more than 1,400 copies of the *B. rapa*, *B. oleracea,* and *B. napus* genomes with high levels of insertion polymorphisms within and among *Brassica* species (Sampath et al. 2013). With the availability of the whole genome sequences of *B. rapa* and *B. oleracea*, about 20 MITE families with approximately 12,000 MITE-like elements were identified in *B. rapa* and *B. oleracea*, 50% of which resided within 2 kb of a gene (Sampath et al. 2014). This will be a valuable resource for the *Brassica* community to understand evolutionary, structural, and functional roles of TEs in *Brassica* species.

7.5 Asymmetric Rates of DNA Removal and Nucleotide Divergence Between *B. rapa* and *B. oleracea*

In plants, unequal homologous recombination (UR) and illegitimate recombination (IR) are two major mechanisms to counteract genome expansion. UR can occur between the 5' and 3' LTR of a single intact element to remove one LTR and the internal part of the element to form a solo LTR (Devos et al. 2002; Ma et al. 2004). Thus, the ratio of solo LTRs to intact LTR-RTs at some extent can reflect the rate of DNA removal in plant genomes (Bennetzen et al. 2005). It was indicated that the ratio of solo LTRs to intact elements in the assembled genome of *B. rapa* is significantly higher than that in the assembled sequences of *B. oleracea*, suggesting that *B. rapa* has a higher rate of DNA removal than *B. oleracea* (Zhao et al. 2013). Thus, it is because the higher frequency of TE accumulation and lower level of DNA elimination of *B. oleracea* that makes its genome larger than *B. rapa*.

The lineage of *B. oleracea* species has been estimated to diverge from the lineage of *B. rapa* species ~3.75 million years (Inaba and Nishio 2002). Recent determination of the reference genomes of *B. rapa* and *B. oleracea* identified 22 orthologous LTR-RT sites and 243 triplicates shared by the two diploid genomes reveals more rapid nucleotide substitution in *B. rapa* than in *B. oleracea*. Further investigation of the triplicated genes retained from the shared whole-genome triplication event indicates that *B. oleracea* has significantly lower rates of synonymous substitution and obviously higher rates of non-synonymous substitution than *B. rapa*, suggesting that the triplicates retained in *B. rapa* have undergone stronger purifying selection than their orthologs in *B. oleracea* (Zhao et al. 2013). These indicate that both the LTR-RTs and genic sequences of the two sequenced diploid genomes evolved asymmetrically after diverging from the common ancestor.

The interspecific asymmetric evolution was perhaps affected by the variation of recombination rates of *B. rapa* and *B. oleracea*. Previous investigation in rice (Tian et al. 2009) and soybean (Du et al. 2012) indicate that the removal of LTR-RT sequences by UR may be associated with genetic recombination. Overall, a higher rate of genetic recombination was observed in *B. rapa* than that in *B. oleracea*, which may partially explain the causes that have driven the asymmetric evolution between these two *Brassica* genomes.

7.6 TE Variation During Polyploidization

Polyploidy, including autopolyploidy (whole-genome duplication of the same individual) and allopolyploidy (hybridization of two different species), is prominent in plants. It is suggested that over 70% of all angiosperm species are polyploids or show evidence of paleopolyploidy (Meyers and Levin 2006; Otto 2007; Jiao et al. 2011). During polyploidization, genome reorganization is observed, such as DNA removal and gene loss, TE amplification or elimination, and chromosomal rearrangements (Zhao et al. 1998; Pires et al. 2004; Udall et al. 2005; Petit et al. 2010;Parisod and Senerchia 2012). Following the 'genome shock' hypothesis proposed by Barbara McClintock (1984), TE activities could be induced by interspecific hybridization or allopolyploidization, probably caused by DNA demethylation which leads to the reactivation of previously silenced TEs (Madlung and Comai 2004). *B. napus* is a natural allotetroploid between the two *Brassica* diploid species *B. rapa* and *B. oleracea*. Previous investigation of three contrasting TE families (*Athila*-like retrotransposon, *BraSto*, and *Bot1*) between resynthesized *B. napus* allotetroploids and their diploid progenitors using sequence-specific amplification polymorphism markers only detected several transposition events limited to some young TEs from the *B. oleracea* genome and did not observe a large burst of transposition and massive structural changes associated with the three TE families (Sarilar et al. 2013). Accordingly, although it was indicated that *Bot1* has amplified in *B. oleracea*, no evidence of a specific burst of *Bot1* in polyploids was detected (Alix and Heslop-Harrison 2004; Alix et al. 2008). However, another study suggests that chromosomal rearrangements and retrotransposon activations were observed extensively in the early generations of interspecific hybridization between *B. napus* and *B. rapa* (Zou et al. 2011). Interestingly, investigation of different types of TEs in different generations of the resynthesized allotetroploids reveals that the transposition frequency was detected at the highest level at the F_2 generation, and varied with the orders of TEs during the process of allopolyploidization (An et al. 2013). Although polyploidy seems to induce the transcriptional activation of some specific TEs, it is still far from understanding the mechanism and the real scenario of TEs involved in polyploidy, no matter in *Brassica* or any other plant species.

7.7 Potential Effects of TEs on Gene Regulation

TEs were designated as 'controlling elements' by Barbara McClintock because of their capacity to regulate gene expression (McClintock 1956a, b). The primary activity of TEs is their inactivation of genes by insertions into the coding sequences or attenuation of gene expression when insertions occur on untranslated or upstream regions or introns. For instance, transposon insertions cause several genes required for color production loss of function, such as a CACTA element in a soybean *flavonoid 3'-hydroxylase* (*F3'H*) gene (Zabala and Vodkin 2005, 2008), a DNA transposon in the transcriptional regulator *bHLH2* of the common morning glory (Clegg and Durbin 2000; Park et al. 2007), and *Mu* elements in the *Bronze-1* gene in maize (Taylor et al. 1986). TE insertion polymorphisms were found to be extremely interesting in the four alleles of the *FATTY ACID ELONGATION1* gene in yellow mustard (*Sinapis alba*) of Brassicaceae, with one null allele harboring a *PIF/Harbinger*-like DNA TE in the coding sequence, another two having a *Copia*-like retrotransposon insertion in the 5' UTR (untranslated region), but with the epigenetic difference in the promoter region, which caused the expression differentiation (Zeng and Cheng 2014).

Given that *A. thaliana* has a much smaller genome, it was estimated that over 1.2% of the expressed, annotated genes contain TE-related fragments, with an underrepresentation of *Helitrons*, and an overrepresentation of *Copia*-like and CACTA-like sequences relative to their current abundance (Lockton and Gaut 2009). Complementally, another research reveals that 3% of TEs in *A. thaliana* are located within gene bodies, particularly in intronic sequences (Le et al. 2015). While in the larger *Brassica* species, profiling of transcriptional activities of TEs using EST and cDNA sequences demonstrates that 6.1% genes were interrupted by LTR-RTs, and also suggests that transcriptional activation and preferential insertions varied with different types of TEs (Gao et al. 2012). Furthermore, a whole-genome-level survey on gene disruption of TE insertions in *B. rapa* and *B. oleracea* using orthologous genes in *A. thaliana* detected 337 genes harboring TEs, of which, 65% contain the insertions in introns (Fig. 7.2; Zhao et al. 2013). To some extent, this is due to the effects of selection that TE insertions in exons are deleterious and would be rapidly removed from the population, leaving high proportion of TEs locating in introns. Theoretically, TE insertions in introns could be removed by splicing. However, large insertions may lead to inefficient or inaccurate recognition of exon-intron boundaries, thus produce premature or non-sense products, or novel proteins that have a distinct function from the original forms. Meanwhile, as most of the TEs are heavily methylated, the effects of epigenetic silencing of TEs were also observed to dilute the expression of host genes even the insertions are located in intronic sequences (Liu et al. 2004). Interestingly, class II elements were found to have a higher frequency to insert into genes than class I retrotransposons in *B. rapa* and *B. oleracea* (Zhao et al. 2013), probably because DNA TEs are generally shorter, and are localized in the intergenic regions of chromosomal arms.

Given that TEs often contain their own promoters, enhancers, and other requisite transcriptional components, it is believed that they have potential effects to regulate flanking genes (Fernandez et al. 2010; Rebollo et al. 2012; Bennetzen and Wang 2014). It was reported that Wis 2-1A LTR-RTs drive the readout synthesis of new transcripts from adjacent sequences including the sense or antisense strands of known genes in newly synthesized wheat amphiploids (Kashkush et al. 2002, 2003). In the two *Brassica* diploid species *B. rapa* and *B. oleracea*, 37 and 70 LTR-RTs were found to have driven the readout transcripts from the 3' LTRs into their adjacent sequences with tissue polymorphisms (Zhao et al. 2013). These readout transcripts could generate double-stranded RNAs (dsRNAs) with the downstream genes which transcribe in different orientations, thus trigger the production of small interfering RNAs (siRNAs) to regulate

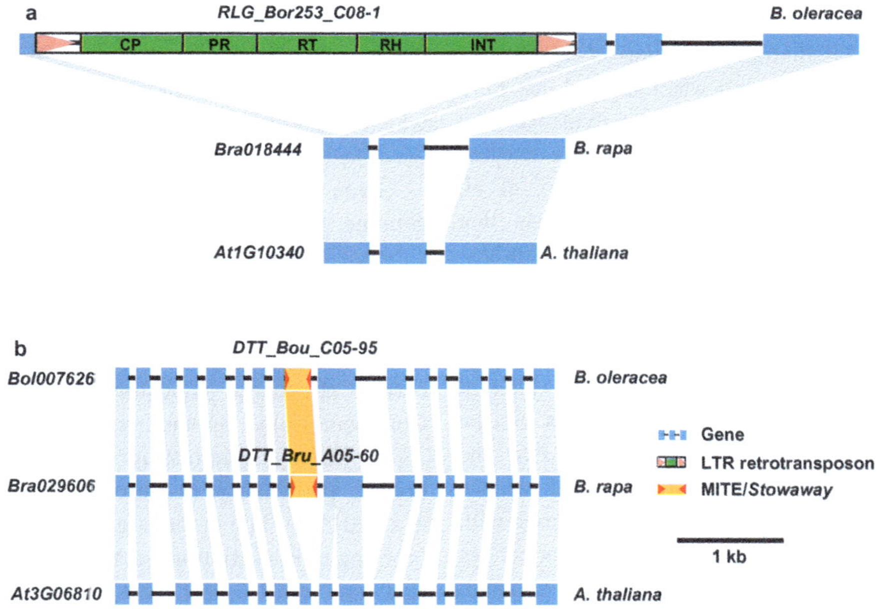

Fig. 7.2 Two examples of TE insertions within gene bodies. **a** A specific LTR-RT insertion in the first exon of the gene in *B. oleracea*. CP, capsid-like protein; PR, protease; INT, integrase; RT, reverse transcriptase; RH, RNAase-H. **b** A shared MITE/*Stowaway* insertion in the introns of the orthologous genes in *B. oleracea* and *B. rapa*. The example data used here was selected from Zhao et al. (2013)

corresponding TEs and genes in *cis* or in *trans* (Lisch 2009, 2013; Zhao and Ma 2013). Hypermethylated TEs may have deleterious effects on their adjacent genes, resulting in down-regulation of neighboring gene expression (Hollister and Gaut 2009; Woodhouse et al. 2014). In addition, TE insertions near or far from genes may function as regulatory components of gene promoters or enhancers long distantly affect gene expression. For example, some TE families acquired the consensus sequence of the E2F transcription factor binding site and amplified it to high copies in *Brassica* species. Interestingly, these consensus sequences in TEs can bind to the E2F transcription factor in vivo, suggesting their potential roles in the E2F transcriptional network (Henaff et al. 2014). To date, limited is known with respect to the molecular mechanisms that how some TEs escaped silencing to regulate genes; probably epigenetic pathways are largely involved.

7.8 Epigenetic Regulation of TEs in *Brassica* Genomes

In addition to UR and IR to remove TE DNA, the genome has evolved epigenetic mechanisms to suppress the activity of TEs. These include small interfering RNAs (siRNAs), DNA cytosine methylation, and histone lysine modification (Lippman and Martienssen 2004; Kasschau et al. 2007; Slotkin and Martienssen 2007; Zilberman et al. 2007). In plants, dsRNAs are processed into two types of siRNAs to trigger epigenetic silencing. Posttranscriptional gene silencing (PTGS) requires 21–22 nucleotide (nt) siRNAs, whereas transcriptional gene silencing (TGS) or RNA-directed DNA methylation (RdDM) is generally triggered by 24–26 nt siRNAs derived from TEs or other tandem repeats (Aufsatz et al. 2002; Hamilton et al. 2002). Although RdDM and TE silencing have been studied extensively in the model Brassicaceae species *A. thaliana*, few investigations have been performed on *Brassica* species. Nevertheless, a study on epigenetic chromatin modifications in *Brassica* genomes reveals that the patterns of DNA and histone H3 methylation differ between the two closely related *Brassica* species *B. rapa* and *B. oleracea*. It shows that DNA methylation was detected in both euchromatic and heterochromatic regions in *B. oleracea* and *B. napus*, but was only enriched exclusively in the heterochromatin in *B. rapa*, suggesting that chromosome structures are largely different between these two diploids (Braszewska-Zalewska et al. 2010). Furthermore, epigenetic changes, including alteration of cytosine methylation and reactivation of silenced TEs, were found to be frequent in the early generations of synthetic *B. napus* allopolyploids (Song et al. 1995; Liu and Wendel 2003; Lukens et al. 2006; Gaeta et al. 2007). This may indicate interspecific hybridization relieves the suppression of TEs, perhaps through the demethylation epigenetic pathway. Comparative analysis of F1 and F7 generations of two resynthesized allotetraploid lines and nature allotetraploid *Arabidopsis suecica* with their two parents *A. thaliana* and *Arabidopsis arenosa* shows dramatic changes in small RNA expression. This study also demonstrates that small RNAs produced during polyploidization or interspecific hybridization may serve as a buffer against the genome shock in allopolyploids and interspecific hybrids (Ha et al. 2009). Polyploidy may provide a common relief on the suppression of TEs, meanwhile induces the reshuffling of epigenetic marks.

7.9 Conclusions and Perspectives

TEs are the major DNA component of the *B. oleracea* genome, and explain approximately 60% of the difference in genome size compared to the closely related *B. rapa* genome. The distribution, amplification, elimination, and evolutionary rates of TEs also show a dramatic difference between these two diploids. TEs could cause mutagenesis of genes or attenuation of gene expression. In addition, they were found to be functional as regulatory elements of gene promoters or enhancers long distantly affect gene expression.

Despite a growing number of studies demonstrating the significance of TEs on genes or during genome evolution in *B. oleracea* and other *Brassica* species, many questions remain unanswered. For example, what are the evolutionary forces or the molecular mechanisms shaping the genome differentiation between *B. rapa* and *B. oleracea*? How important are TEs responsible for *Brassica* genome evolution or speciation? Additionally, what are responses of TEs to allopolyploidy or interspecific hybridization? Moreover, although TEs are extensively suppressed by epigenetic pathways, the epigenetic regulation of TEs is largely unclear in *Brassica*. The availability of the reference sequences of *B. rapa*, *B. oleracea*, *B. napus*, and other ongoing sequenced *Brassica* species will facilitate addressing these questions in future work.

Acknowledgements We apologize to our colleagues whose contributions have not been cited or discussed here due to space limitations. This work was supported by the National Key Research Program to S.L., by Purdue Agricultural Research Program to J.M., and by Miami University Startup Funds to M.Z.

References

Alix K, Heslop-Harrison JS (2004) The diversity of retroelements in diploid and allotetraploid *Brassica* species. Plant Mol Biol 54:895–909

Alix K, Ryder CD, Moore J, King GJ, Heslop-Harrison JS (2005) The genomic organization of retrotransposons in *Brassica oleracea*. Plant Mol Biol 59:839–851

Alix K, Joets J, Ryder CD, Moore J, Barker GC, Bailey JP, King GJ, Heslop-Harrison JS (2008) The CACTA transposon *Bot1* played a major role in *Brassica* genome divergence and gene proliferation. Plant J 56:1030–1044

An Z, Tang Z, Ma B, Mason AS, Guo Y, Yin J, Gao C, Wei L, Li J, Fu D (2014) Transposon variation by order during allopolyploidisation between *Brassica oleracea* and *Brassica rapa*. Plant Biol 16:825–835

Arnaud P, Yukawa Y, Lavie L, Pelissier T, Sugiura M, Deragon JM (2001) Analysis of the SINE S1 Pol III promoter from *Brassica*; impact of methylation and influence of external sequences. Plant J 26:295–305

Aufsatz W, Mette MF, van der Winden J, Matzke AJM, Matzke M (2002) RNA-directed DNA methylation in *Arabidopsis*. Proc Natl Acad Sci USA 99:16499–16506

Beilstein MA, Al-Shehbaz IA, Kellogg EA (2006) Brassicaceae phylogeny and trichome evolution. Amer J Bot 93:607–619

Bennetzen JL, Wang H (2014) The contributions of transposable elements to the structure, function, and evolution of plant genomes. Annu Rev Plant Biol 65:505–530

Bennetzen JL, Ma J, Devos KM (2005) Mechanisms of recent genome size variation in flowering plants. Ann Bot 95:127–132

Braszewska-Zalewska A, Bernas T, Maluszynska J (2010) Epigenetic chromatin modifications in *Brassica* genomes. Genome 53:203–210

Bureau TE, Wessler SR (1992) Tourist: a large family of small inverted repeat elements frequently associated with maize genes. Plant Cell 4:1283–1294

Bureau TE, Wessler SR (1994) Stowaway: a new family of inverted repeat elements associated with the genes of both monocotyledonous and dicotyledonous. plants Plant Cell 6:907–916

ChalhoubB Denoeud F, Liu S, Parkin IA, Tang H et al (2014) Early allopolyploid evolution in the post-Neolithic *Brassica napus* oilseed genome. Science 345:950–953

Cheung F, Trick M, Drou N, Lim YP, Park JY et al (2009) Comparative analysis between homoeologous genome segments of *Brassica napus* and its progenitor species reveals extensive sequence-level divergence. Plant Cell 21:1912–1928

Clegg MT, Durbin, ML (2000) Flower color variation: a model for the experimental study of evolution. Proc Natl Acad Sci USA 97:7016–7023

Deragon JM, Zhang X (2006) Short interspersed elements (SINEs) in plants: origin, classification, and use as phylogenetic markers. Syst Biol 55:949–956

Deragon JM, Gilbert N, Rouquet L, Lenoir A, Arnaud P, Picard G (1996) A transcriptional analysis of the S1Bn (*Brassica napus*) family of SINE retroposons. Plant Mol Biol 32:869–878

Deragon JM, Landry BS, Pelissier T, Tutois S, Tourmente S, Picard G (1994) An analysis of retroposition in plants based on a family of SINEs from *Brassica napus*. J Mol Evol 39:378–386

Devos KM, Brown JKM, Bennetzen JL (2002) Genome size reduction through illegitimate recombination counteracts genome expansion in *Arabidopsis*. Genome Res 12:1075–1079

Du J, Tian Z, Bowen NJ, Schmutz J, Shoemaker RC, Ma J (2010) Bifurcation and enhancement of autonomous-nonautonomous retrotransposon partnership through LTR swapping in soybean. Plant Cell 22:48–61

Du J, Tian Z, Sui Y, Zhao M, Song Q, Cannon SB, Cregan P, Ma J (2012) Pericentromeric effects shape the patterns of divergence, retention, and expression of duplicated genes in the paleopolyploid soybean. Plant Cell 24:21–32

Fedoroff NV (2012) Transposable elements, epigenetics, and genome evolution. Science 338:758–767
Fernandez L, Torregrosa L, Segura V, Bouquet A, Martinez-Zapater JM (2010) Transposon-induced gene activation as a mechanism generating cluster shape somatic variation in grapevine. Plant J 61:545–557
Finnegan DJ (1989) Eukaryotic transposable elements and genome evolution. Trends Genet 5:103–107
Gaeta RT, Pires JC, Iniguez-Luy F, Leon E, Osborn TC (2007) Genomic changes in resynthesized *Brassica napus* and their effect on gene expression and phenotype. Plant Cell 19:3403–3417
Gao C, Xiao M, Jiang L, Li J, Yin J, Ren X, Qian W, Oscar O, Fu D, Tang Z (2012) Characterization of transcriptional activation and inserted-into-gene preference of various transposable elements in the *Brassica* species. Mol Biol Rep 39:7513–7523
Goubely C, Arnaud P, Tatout C, Heslop-Harrison JS, Deragon JM (1999) S1 SINE retroposons are methylated at symmetrical and non-symmetrical positions in *Brassica napus*: identification of a preferred target site for asymmetrical methylation. Plant Mol Biol 40:197–197
Grandbastien MA (1992) Retroelements in higher-plants. Trends Genet 8:103–108
Ha M, Lu J, Tian L, Ramachandran V, Kasschau K et al (2009) Small RNAs serve as a genetic buffer against genomic shock in *Arabidopsis* interspecific hybrids and allopolyploids. Proc Natl Acad Sci USA 106:17835–17840
Hamilton A, Voinnet O, Chappell L, Baulcombe D (2002) Two classes of short interfering RNA in RNA silencing. EMBO J 21:4671–4679
Harrison GE, Heslopharrison JS (1995) Centromeric repetitive DNA sequences in the genus *Brassica*. Theor Appl Genet 90:157–165
Henaff E, Vives C, Desvoyes B, Chaurasia A, Payet J, Gutierrez C, Casacuberta JM (2014) Extensive amplification of the E2F transcription factor binding sites by transposons during evolution of *Brassica* species. Plant J 77:852–862
Hollister JD, Gaut BS (2007) Population and evolutionary dynamics of *Helitron* transposable elements in *Arabidopsis thaliana*. Mol Biol Evol 11:2515–2524
Hollister JD, Gaut BS (2009) Epigenetic silencing of transposable elements: a trade-off between reduced transposition and deleterious effects on neighboring gene expression. Genome Res 19:1419–1428
Inaba R, Nishio T (2002) Phylogenetic analysis of *Brassiceae* based on the nucleotide sequences of the S-locus related gene, *SLR1*. Theor Appl Genet 105:1159–1165
Jiang J, Birchler JA, Parrott WA, Dawe RK (2003) A molecular view of plant centromeres. Trends Plant Sci 8:570–575
Jiang N, Bao Z, Temnykh S, Cheng Z, Jiang J, Wing RA, McCouch SR, Wessler SR (2002a) Dasheng: a recently amplified nonautonomous long terminal repeat element that is a major component of pericentromeric regions in rice. Genetics 161:1293–1305
Jiang N, Jordan IK, Wessler SR (2002b) Dasheng and RIRE2 A nonautonomous long terminal repeat element and its putative autonomous partner in the rice genome. Plant Physiol 130:1697–1705
Jiang N, Bao Z, Zhang X, Eddy SR, Wessler SR (2004a) Pack-MULE transposable elements mediate gene evolution in plants. Nature 431:569–573
Jiang N, Feschotte C, Zhang X, Wessler SR (2004b) Using rice to understand the origin and amplification of miniature inverted repeat transposable elements (MITEs). Curr Opin Plant Biol 7:115–119
Jiao Y, Wickett NJ, Ayyampalayam S, Chanderbali AS, Landherr L et al (2011) Ancestral polyploidy in seed plants and angiosperms. Nature 473:97–100
Jiao Y, Peluso P, Shi J, Liang T, Stitzer MC et al (2017) Improved maize reference genome with single-molecule technologies. Nature 546:524–527
Johnston JS, Pepper AE, Hall AE, Chen ZJ, Hodnett G, Drabek J, Lopez R, Price HJ (2005) Evolution of genome size in Brassicaceae. Ann Bot 95:229–235
Kalendar R, Vicient CM, Peleg O, Anamthawat-Jonsson K, Bolshoy A, Schulman AH (2004) Large retrotransposon derivatives: abundant, conserved but nonautonomous retroelements of barley and related genomes. Genetics 166:1437–1450
Kapitonov VV, Jurka J (2001) Rolling-circle transposons in eukaryotes. Proc Natl Acad Sci USA 98:8714–8719
Kapitonov VV, Jurka J (2006) Self-synthesizing DNA transposons in eukaryotes. Proc Natl Acad Sci USA 103:4540–4545
Kashkush K, Feldman M, Levy AA (2002) Gene loss, silencing and activation in a newly synthesized wheat allotetraploid. Genetics 160:1651–1659
Kashkush K, Feldman M, Levy AA (2003) Transcriptional activation of retrotransposons alters the expression of adjacent genes in wheat. Nat Genet 33:102–106
Kasschau KD, Fahlgren N, Chapman EJ, Sullivan CM, Cumbie JS, Givan SA, Carrington JC (2007) Genome-wide profiling and analysis of *Arabidopsis* siRNAs. PLoS Biol 5:479–493
Kejnovsky E, Kubat Z, Macas J, Hobza R, Mracek J, Vyskot B (2006) Retand: a novel family of gypsy-like retrotransposons harboring an amplified tandem repeat. Mol Genet Genomics 276:254–263
Kumar A, Bennetzen JL (1999) Plant retrotransposons. Annu Rev Genet 33:479–532
Kwon SJ, Kim DH, Lim MH, Long Y, Meng JL et al (2007) Terminal repeat retrotransposon in miniature (TRIM) as DNA markers in *Brassica* relatives. Mol Genet Genomics 278:361–370
Lander ES, Linton LM, Birren B, Nusbaum C, Zody MC et al (2001) Initial sequencing and analysis of the human genome. Nature 409:860–921

Lippman Z, Martienssen R (2004) The role of RNA interference in heterochromatic silencing. Nature 431:364–370

Le TN, Miyazaki Y, Takuno S, Saze H (2015) Epigenetic regulation of intragenic transposable elements impacts gene transcription in *Arabidopsis thaliana*. Nucleic Acids Res 43:3911–3921

Lenoir A, Pelissier T, Bousquet-Antonelli C, Deragon JM (2005) Comparative evolution history of SINEs in *Arabidopsis thaliana* and *Brassica oleracea*: evidence for a high rate of SINE loss. Cytogenet Genome Res 110:441–447

Lenoir A, Cournoyer B, Warwick S, Picard G, Deragon JM (1997) Evolution of SINE S1 retroposons in Cruciferae plant species. Mol Biol Evol 14:934–941

Lim KB, de Jong H, Yang TJ, Park JY, Kwon SJ et al (2005) Characterization of rDNAs and tandem repeats in the heterochromatin of *Brassica rapa*. Mol Cells 19:436–444

Lim KB, Yang TJ, Hwang YJ, Kim JS, Park YJ et al (2007) Characterization of the centromere and pericentromere retrotransposons in *Brassica rapa* and their distribution in related *Brassica* species. Plant J 49:765–765

Lisch D (2009) Epigenetic regulation of transposable elements in plants. Annu Rev Plant Biol 60:43–66

Lisch D (2013) How important are transposons for plant evolution? Nat Rev Genet 14:49–61

Liu B, Wendel JF (2003) Epigenetic phenomena and the evolution of plant allopolyploids. Mol Phylogenet Evol 29:365–379

Liu J, He Y, Amasino R, Chen X (2004) siRNAs targeting an intronic transposon in the regulation of natural flowering behavior in *Arabidopsis*. Genes Dev 18:2873–2878

Liu S, Liu Y, Yang X, Tong C, Edwards D et al (2014) The *Brassica oleracea* genome reveals the asymmetrical evolution of polyploidy genomes. Nat Commun 5:3930

Lockton S, Gaut BS (2009) The contribution of transposable elements to expressed coding sequence in *Arabidopsis thaliana* J Mol Evol 68:80–89

Lukens LN, Pires JC, Leon E, Vogelzang R, Oslach L, Osborn T (2006) Patterns of sequence loss and cytosine methylation within a population of newly resynthesized *Brassica napus*allopolyploids. Plant Physiol 140:336–348

Ma J, Devos KM, Bennetzen JL (2004) Analyses of LTR-retrotransposon structures reveal recent and rapid genomic DNA loss in rice. Genome Res 14:860–869

Ma J, Wing RA, Bennetzen JL, Jackson SA (2007) Plant centromere organization: a dynamic structure with conserved functions. Trends Genet 23:134–139

Madlung A, Comai L (2004) The effect of stress on genome regulation and structure. Ann Bot 94:481–495

McClintock B (1956a) Intranuclear systems controlling gene action and mutation. Brookhaven Symp Biol 8:58–74

McClintock B (1956b) Controlling elements and the gene. Brookhaven Symp Biol 21:197–216

McClintock B (1984) The significance of responses of the genome to challenge. Science 226:792–801

Meyers LA, Levin DA (2006) On the abundance of polyploids in flowering plants. Evolution 60:1198–1206

Murukarthick J, Sampath P, Lee SC, Choi BS, Senthil N, Liu S, Yang TJ (2014) BrassicaTED-a public database for utilization of miniature transposable elements in *Brassica* species. BMC Res Notes 7:379

Myouga F, Tsuchimoto S, Noma K, Ohtsubo H, Ohtsubo H (2001) Identification and structural analysis of SINE elements in the *Arabidopsis thaliana* genome. Genes Genet Syst 76:169–179

Nagaki K, Cheng ZK, Ouyang S, Talbert PB, Kim M, Jones KM, Henikoff S, Buell CR, Jiang J (2004) Sequencing of a rice centromere uncovers active genes. Nat Genet 36:138–145

Nouroz F, Noreen S, Heslop-Harrison JS (2015) Evolutionary genomics of miniature inverted-repeat transposable elements (MITEs) in *Brassica*. Mol Genet Genomics 290:2297–2312

Otto SP (2007) The evolutionary consequences of polyploidy. Cell 131:452–462

Park KI, Ishikawa N, Morita Y, Choi JD, Hoshino A, Iida S (2007) A bHLH regulatory gene in the common morning glory, *Ipomoea purpurea*, controls anthocyanin biosynthesis in flowers, proanthocyanidin and phytomelanin pigmentation in seeds, and seed trichome formation. Plant J49:641–654

Parisod C, Senerchia N (2012) The response of transposable elements to polyploidy. In: Grandbastien MA, Casacuberta JM (eds) Plant transposable elements. Topics in Current Genetics Springer, Berlin, Heidelberg, pp 147–168

Petit M, Guidat C, Daniel J, Denis E, Montoriol E et al (2010) Mobilization of retrotransposons in synthetic allotetraploid tobacco. New Phytol 186:135–147

Pires JC, Zhao JW, Schranz ME, Leon EJ, Quijada PA, Lukens LN, Osborn TC (2004) Flowering time divergence and genomic rearrangements in resynthesized *Brassica* polyploids (Brassicaceae). Biol J Linn Soc 82:675–688

Pritham EJ, Feschotte C (2007) Massive amplification of rolling-circle transposons in the lineage of the bat *Myotis lucifugus*. Proc Natl Acad Sci USA 6:895–1900

Qiu D, Gao MQ, Li GY, Quiros C (2009) Comparative sequence analysis for *Brassica oleracea* with similar sequences in *B rapa* and *Arabidopsis thaliana*. Plant Cell Rep 28:649–661

Rebollo R, Romanish MT, Mager DL (2012) Transposable elements: an abundant and natural source of regulatory sequences for host genes. Annu Rev Genet 46:21–42

Sampath P, Lee SC, Lee J, Izzah NK, Choi BS, Jin M, Park BS, Yang TJ (2013) Characterization of a new high copy Stowaway family MITE, *BRAMI*-1 in *Brassica* genome. BMC Plant Biol 13:56

Sampath P, Murukarthick J, Izzah NK, Lee J, Choi HI, Shirasawa K, Choi BS, Liu SY, Nou IS, Yang TJ (2014) Genome-wide comparative analysis of 20 miniature inverted-repeat transposable element families in *Brassica rapa* and *B.oleracea*. PLoS One 9:e94499

Sarilar V, Marmagne A, Brabant P, Joets J, Alix K (2011) BraSto, a Stowaway MITE from *Brassica*: recently active copies preferentially accumulate in the gene space. Plant Mol Biol 77:59–75

Sarilar V, Palacios PM, Rousselet A, Ridel C, Falque M, Eber F, Chevre AM, Joets J, Brabant P, Alix K (2013) Allopolyploidy has a moderate impact on restructuring at three contrasting transposable element insertion sites in resynthesized *Brassica napus*allotetraploids. New Phytol 198:593–604

Schnable PS, Ware D, Fulton RS, Stein JC, Wei F et al (2009) The B73 maize genome: complexity, diversity, and dynamics. Science 326:1112–1115

Slotkin RK, Martienssen R (2007) Transposable elements and the epigenetic regulation of the genome. Nat Rev Genet 8:272–285

Song KM, Lu P, Tang KL, Osborn TC (1995) Rapid genome change in synthetic polyploids of *Brassica* and its implications for polyploid evolution. Proc Natl Acad Sci USA 92:7719–7723

Tatout C, Warwick S, Lenoir A, Deragon JM (1999) SINE insertions as clade markers for wild crucifer species. Mol Biol Evol 16:1614–1621

Taylor LP, Chandler VL, Walbot V (1986) Insertion of 1.4 KB and 1.7 KB*Mu* elements into the *bronze1* gene of *Zea mays* L. Maydica 31:31–45

The Arabidopsis Genome Initiative (2000) Analysis of the genome sequence of the flowering plant *Arabidopsis thaliana*. Nature 408:796–815

Tian Z, Rizzon C, Du J, Zhu L, Bennetzen JL, Jackson SA, Gaut BS, Ma J (2009) Do genetic recombination and gene density shape the pattern of DNA elimination in rice long terminal repeat retrotransposons? Genome Res 19:2221–2230

Udall JA, Quijada PA, Osborn TC (2005) Detection of chromosomal rearrangements derived from homoeologous recombination in four mapping populations of *Brassica napus* L. Genetics 169:967–979

Wang X, Wang H, Wang J, Sun R, Wu J et al (2011a) The genome of the mesopolyploid crop species *Brassica rapa*. Nat Genet 43:1035–1039

Wang G, He Q, Liu F, Cheng Z, Talbert PB, Jin W (2011b) Characterization of CENH3 proteins and centromere-associated DNA sequences in diploid and allotetraploid *Brassica* species. Chromosoma 120:353–365

Wang W, Huang S, Liu Y, Fang Z, Yang L, et al (2012) Construction and analysis of a high-density genetic linkage map in cabbage (Brassica oleracea L. var. capitata). BMC Genomics 13:523

Wicker T, Sabot F, Hua-Van A, Bennetzen JL, Capy P et al (2007) A unified classification system for eukaryotic transposable elements. Nat Rev Genet 8:973–982

Witte CP, Le QH, Bureau T, Kumar A (2001) Terminal-repeat retrotransposons in miniature (TRIM) are involved in restructuring plant genomes. Proc Natl Acad Sci USA 98:13778–13783

Woodhouse MR, Cheng F, Pires JC, Lisch D, Freeling M, Wang X (2014) Origin, inheritance, and gene regulatory consequences of genome dominance in polyploids. Proc Natl Acad Sci USA 111:5283–5288

Yang J, Liu D, Wang X, Ji C, Cheng F et al (2016) The genome sequence of allopolyploid *Brassica juncea* and analysis of differential homoeolog gene expression influencing selection. Nat Genet 48:1225–1232

Yang TJ, Kwon SJ, Choi BS, Kim JS, Jin M et al (2007) Characterization of terminal-repeat retrotransposon in miniature (TRIM) in *Brassica* relatives. Theor Appl Genet 114:627–636

Yang TJ, Kim JS, Kwon SJ, Lim KB, Choi BS et al (2006) Sequence-level analysis of the diploidization process in the triplicated FLOWERING LOCUS C region of *Brassica rapa*. Plant Cell 18:1339–1347

Yang YW, Lai KN, Tai PY, Li WH (1999) Rates of nucleotide substitution in angiosperm mitochondrial DNA sequences and dates of divergence between *Brassica* and other angiosperm lineages. J Mol Evol 48:597–604

Zabala G, Vodkin LO (2005) The wp mutation of Glycine max carries a gene-fragment-rich transposon of the CACTA superfamily. Plant Cell 17:2619–2632

Zabala G, Vodkin L (2008) A putative autonomous 205 kb-CACTA transposon insertion in an F3'H allele identifies a new CACTA transposon subfamily in *Glycine max*. BMC Plant Biol 8:124

Zeng FQ, Cheng BF (2014) Transposable element insertion and epigenetic modification cause the multiallelic variation in the expression of *FAE1* in *Sinapis alba*. Plant Cell 26:2648–2659

Zhang X, Wessler SR (2004) Genome-wide comparative analysis of the transposable elements in the related species *Arabidopsis thaliana* and *Brassica oleracea*. Proc Natl Acad Sci USA 101:5589–5594

Zhao M, Ma J (2013) Co-evolution of plant LTR-retrotransposons and their host genomes. Protein Cell 4:493–501

Zhao M, Du J, Lin F, Tong C, Yu J, Huang S, Wang X, Liu S, Ma J (2013) Shifts in evolutionary rate and intensity of purifying selection between two *Brassica* genomes revealed by analyses of orthologous transposons and relics of a whole genome triplication. Plant J 76:211–220

Zhao XP, Si Y, Hanson RE, Crane CF, Price HJ, Stelly DM, Wendel JF, Paterson AH (1998) Dispersed

repetitive DNA has colonized new genomes since polyploid formation in cotton. Genome Res 8:479–492
Zhong CX, Marshall JB, Topp C, Mroczek R, Kato A, Nagaki K, Birchler JA, Jiang JM, Dawe RK (2002) Centromeric retroelements and satellites interact with maize kinetochore protein CENH3. Plant Cell 14:2825–2836
Zilberman D, Gehring M, Tran RK, Ballinger T, Henikoff S (2007) Genome-wide analysis of *Arabidopsis thaliana* DNA methylation uncovers an interdependence between methylation and transcription. Nat Genet 39:61–69
Zou J, Fu DH, Gong HH, Qian W, Xia W et al (2011) *De novo* genetic variation associated with retrotransposon activation, genomic rearrangements and trait variation in a recombinant inbred line population of *Brassica napus* derived from interspecific hybridization with *Brassica rapa*. Plant J 68:212–224

The *CONSTANS*-LIKE (*COL*) Genes in Brassica Plants

8

Li Wang, Jinpeng Wang, Weina Ge, Xiaoming Song, Yuxin Pan, Lan Zhang, Jiaqi Zhang, Tianyu Lei, Yuxian Li, Zhenyi Wang, Min Yuan, and Xiyin Wang

Abstract

Large gene families are anticipated to show large-scale copy number changes from species to species due to relatively more freedom in selective pressure. In paleopolyploid plant species, recursive whole-genome duplication/triplication incurred considerable gene copy number increase and subsequent decrease due to genome instability, with a significant influence on large gene families. Here, we describe copy number changes and evolutionary patterns of Brassica *CONSTANS*-LIKE (*COL*) genes, which regulate flowering time, in the diploid Brassica species *B. oleracea*, its close relative *B. rapa*, and their allotetraploid *B. napus*. We found that the whole-genome triplication which formed the Brassica diploids, and the subsequent whole-genome duplication which formed *B. napus*, were accompanied by large-scale gene loss, and certain lineages of genes were positively selected. Interestingly, genes were often alternatively lost or selected, that is, when one gene was preserved or underwent positive selection pressure, its ortholog was frequently lost. This demonstrates the high degree of flexibility of highly duplicated (paleo-triplicated) genomes, particularly exemplified by large gene families.

L. Wang · J. Wang · W. Ge · X. Song · Y. Pan · L. Zhang · J. Zhang · T. Lei · Y. Li · Z. Wang · M. Yuan · X. Wang (✉)
School of Life Sciences, North China University of Science and Technology, Tangshan 063210, Hebei, China
e-mail: wangxiyin@vip.sina.com

L. Wang · J. Wang · W. Ge · X. Song · Y. Pan · L. Zhang · J. Zhang · T. Lei · Y. Li · Z. Wang · M. Yuan · X. Wang
Center for Genomics and Computational Biology, North China University of Science and Technology, Tangshan 063210, Hebei, China

J. Wang
State Key Laboratory of Systematic and Evolutionary Botany, Institute of Botany, Chinese Academy of Sciences, Beijing 100093, China

J. Wang
University of Chinese Academy of Sciences, Beijing 100049, China

8.1 Introduction

8.1.1 Gene Family Size Varies

Gene families vary considerably in size. There are certain gene families with a small number of gene copies, and others with very large copy numbers of genes. A prominent large gene family is the nucleotide-binding site-leucine-rich repeat (NBS-LRR) disease resistance genes, which often contain hundreds of copies in plant species (Velasco et al. 2010). A large number of gene copies allow a family to produce appreciable numbers of mutants to cope with an ever-changing environment (Yang et al. 2008; Mun

S. Liu et al. (eds.), *The Brassica oleracea Genome*, Compendium of Plant Genomes,
https://doi.org/10.1007/978-3-030-31005-9_8

et al. 2009). Frequently, tandem clustering on chromosomes and the distribution of intragenic repetitive domains makes it possible to produce novel mutations through ectopic recombination (Ratnaparkhe et al. 2011).

8.1.2 Recursive Polyploidization and Gene Duplication

A key feature of plants is that they have been recursively affected by polyploidization (Jiao et al. 2012), resulting in duplication/triplication of nuclear genomes, providing a good opportunity to assess the plasticity of a plant genome (Olaitan et al. 2015). Increased copies of genes and chromosomes in which they reside lead to elevated genome instability and often enormous numbers of gene copies were deleted or translocated (Wang et al. 2013). Chromosomes may be rearranged to produce neo-chromosomes (Wang et al. 2016b), and eventually, a resultant neo-genome will regain stability. The polyploidization and subsequent process have a considerable effect on different gene families. Large gene families may increase or decrease their copy number, a process that provides considerable opportunities to mutate and evolve new functions.

The ancestral Brassica diploid underwent a whole-genome triplication after its divergence from Arabidopsis, leading to the evolution of the three diploid species *B. rapa*, *B. oleracea,* and *B. nigra*. In the hybrid between *B. rapa* and *B. oleracea*, *B. napus,* the addition of two very similar ancestral genomes gave rise to an allotetraploid which duplication of most genes present in the two diploid ancestors (Wang et al. 2011b; Chalhoub et al. 2014; Liu et al. 2014). Gene collinearity is pivotal to understand the genome structure and evolution of genomes (Tang et al. 2008; Lee et al. 2013). Actually, near all sequenced genomes employed gene collinearity to deconvolute their complex genome structure and unfold their complex history (Proost et al. 2009; Wang et al. 2016a). Collinear genes are those genes that descend from the same ancestral gene and have preserved the ancestral chromosome locations in extant genomes. Actually, though enormous gene losses occurred, often hundreds of collinear genes can be found in extant genomes. Collinear genes provide information to infer credible gene orthology between genomes or paralogy within or also between genomes, therefore contributing to understand the formation of gene families (Wang et al. 2018).

8.1.3 *CONSTANS*-LIKE (*COL*) Genes

CONSTANS-LIKE (*COL*) comprise a large gene family, with 24 *COL* genes being identified in Arabidopsis. They have been shown to regulate flowering time in Arabidopsis (Fornara et al. 2010). The *COL* genes delay flowering, while *FLOWERING LOCUST* (FT)-like genes promote flowering in short-day plants (Putterill et al. 2004; Song et al. 2015). *COL* genes have been demonstrated to control the timing of the formation of the *FLAVIN-BINDING, KELCH REPEAT, F-BOX 1* (*FKF1*), and *GIGANTEA* (*GI*) gene complexes which determines the timing of daytime *COL* gene expression in long-day plants (Wang et al. 2015b).

Through phylogenetic, evolutionary, and gene-collinearity analysis, we describe here whether and how the Brassica whole-genome triplication and the *B. napus* whole-genome duplication have affected the copy number changes and evolutionary patterns of the *COL* genes among diploid *B. oleracea* and *B rapa* and their allotetraploid *B. napus*.

8.2 Sequence Alignments and Phylogenetic Analyses

8.2.1 Data Sources

Genome sequences from *Arabidopsis thaliana*, *B. oleracea*, *B. rapa*, and *B. napus* were used. The *A. thaliana* sequences were downloaded from TAIR, *B. rapa* sequences from BRAD (Cheng et al. 2011), and the *B. oleracea* sequences were from EMBL (http://www.ebi.ac.

uk/). Sequences of two other representative dicot species (*Vitus vinifera* and *Curcubita pepo*) were downloaded from Phytozome (Goodstein et al. 2012).

8.2.2 Alignment and Tree Construction

Multiple sequence alignment was performed by using the software MUSCLE with default parameters (Edgar 2004), and phylogeny was constructed using previously reported methods (Li et al. 2015). For small groups of genes, neighbor-joining trees were constructed using MEGA 6.0 with 1000 bootstraps and a Poisson correction model (Tamura et al. 2013). For large groups of genes, PhyML 3.0 was employed to construct maximal likelihood trees with the Jones, Taylor, and Thorton (JTT) model, using 100 nonparametric bootstrap replicates (Guindon et al. 2010).

8.3 Identification and Analysis of *COL* Genes

8.3.1 Gene Numbers in Different Plants

Using the 24 Arabidopsis *COL* genes as seeds, we systematically identified *COL* genes in *B. rapa*, *B. oleracea*, and *B. napus*. After filtering out possible pseudogenes with incomplete gene structures, we identified a total of 194 *COL* genes among all species. *B. napus* contained the most *COL* genes (68), very close to the sum of its two diploid progenitors *B. rapa* (34) and *B. oleracea* (36).

Assuming that no gene was lost after polyploidization, any given *A. thaliana* gene would be expected to have three corresponding *B. oleracea* and *B. rapa* orthologs along with six *B. napus* orthologs. However, we found that the Brassica gene numbers are much smaller than expected, with three times the number found in Arabidopsis, suggesting that there was a high possibility of gene loss after whole-genome triplication in the Brassica diploids. A correspondence of approximately 1:1 between the two subgenomes of *B. napus* showed that few *COL*genes appear to have been lost after allotetraploidization.

8.3.2 Phylogenetic Analysis

To understand more about the evolution and functional innovation of *COL* genes, we identified the homologs from the representative dicot species *V. vinifera* and *C. pepo*, respectively. The total number of *COL* genes across all six investigated species added up to 194. To conduct the classification of *COL* genes, we constructed a phylogenetic tree using all 194*COL* genes by using MEGA 6.0 (Fig. 8.1). The *COL* genes could be grouped into nine groups (numbered 1–9). Each group shares genes from all of the investigated species, indicating that they formed prior to the divergence of these dicot plants.

8.3.3 Gene Collinearity Shows Genome Instability

Gene collinearity adds information to identify orthologs between plants or polyploidy-derived paralogs. By referring to gene collinearity, we found variable gene losses during evolution (Fig. 8.2). A check of cross-genome gene collinearity, performed here by studying collinearity-supported chromosomal regions, showed that the *COL* genes have been affected by both polyploidization and tandem duplication. Seven cases of tandemly duplicated genes were located in the different species, revealing the ancestry of the origination of these tandem pairs. In some cases, gene collinearity suggested significant instability of the studied genomic regions with gene numbers in investigated chromosomal regions showing vast differences.

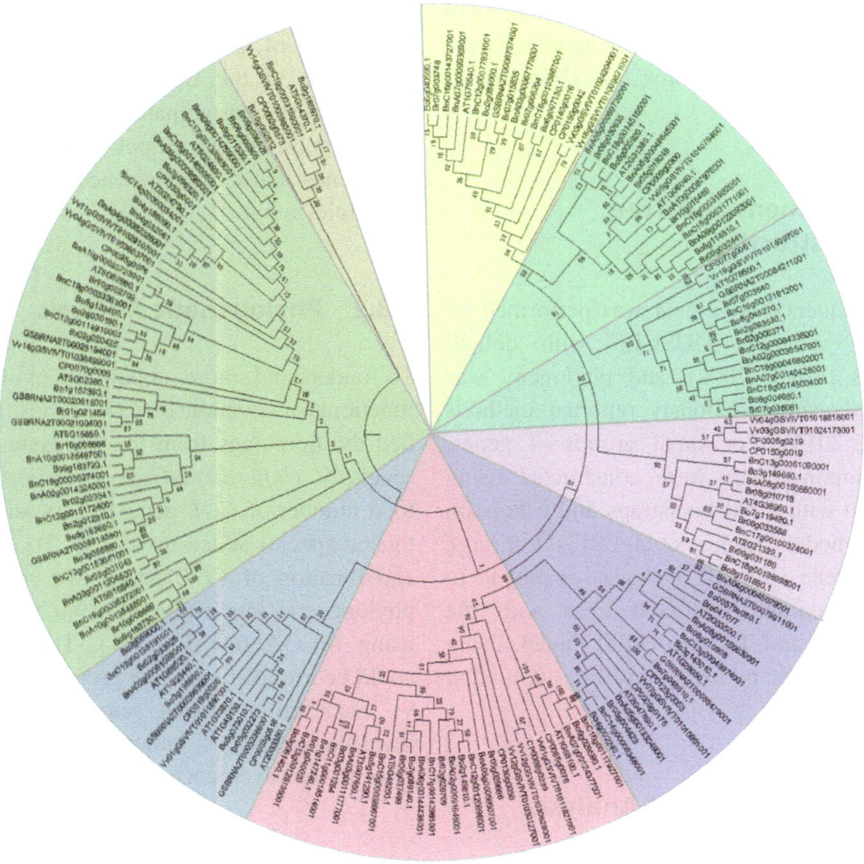

Fig. 8.1 Phylogeny of COL genes from *B. oleracea* compared to *B. rapa*, *B. napus*, *A. thaliana*, *V. viniferus,* and *C. pepo*

8.3.4 Gene Duplication and Gene Loss

The *COL* gene groups were investigated in considerable detail with regard to gene loss. In group 1, the Arabidopsis gene AT1G75540.1 was found to have only one ortholog in each of the investigated diploid Brassica species, namely Bo6g040590.1 in *B. oleracea* and Br07g003748 in *B. rapa*, and the consequently expected two copies BnA07g00099369001 and BnC16g00143727001 in *B. napus*. This is an example of the loss of two expected copies in the two diploid *Brassica* species after the whole-genome triplication. Actually, gene phylogeny shows an alternative correspondence between the diploid genes and the two tetraploid genes (Fig. 8.3). This suggests that the gene losses occurred both after the Brassica-common whole-genome triplication and after the formation of *B. napus* through allotetraploidization. Similar gene losses were also seen in other groups of *COL* genes. For example, in group 2, AT2G31380.1 has all orthologs lost in *B. oleracea*, with only a single copy preserved in *B. rapa* (Br05g018249) and retained in *B. napus* (BnA05g00048645001). In contrast, its paralog, AT1G06040.1, has all orthologs preserved in *B. rapa* and *B. napus*, while only one *B. oleracea* gene was not found, either due to gene loss in *B. oleracea* or incompleteness of the genome assembly.

In general, after the Brassica paleotriplication, there have been considerable gene losses but no gene duplication. We checked 12 subtrees in each of the 8 groups of *COL* genes (Fig. 8.3), each of which has an Arabidopsis outgroup and

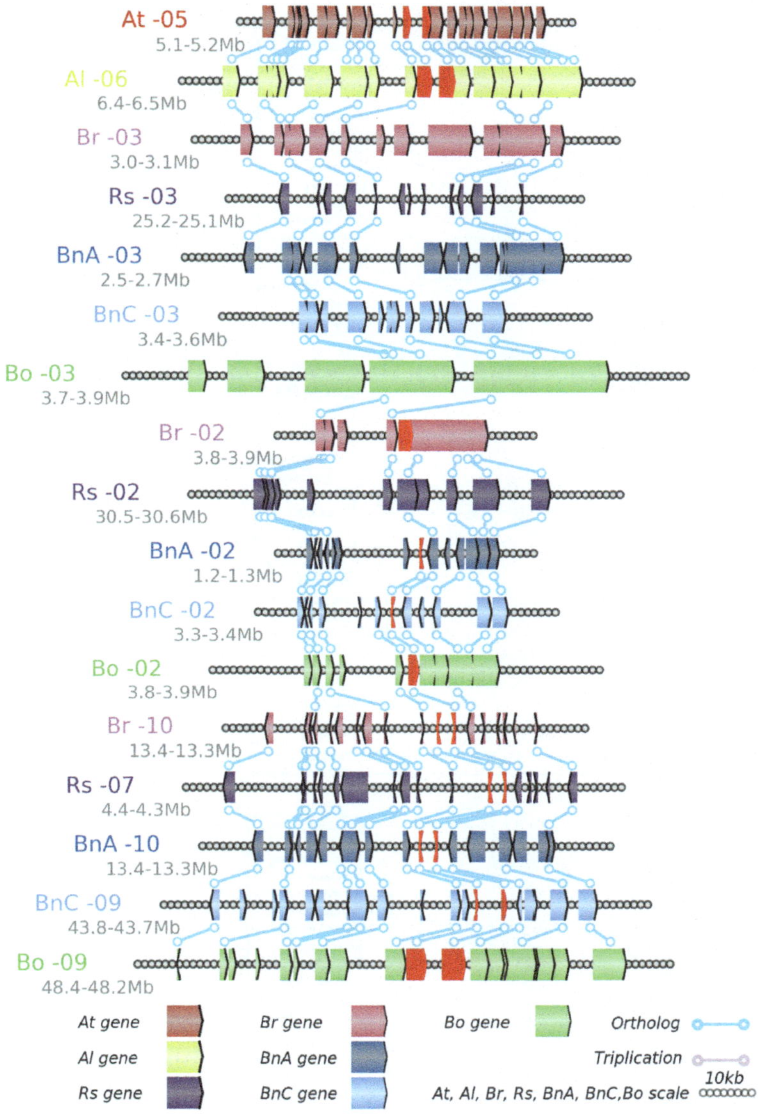

Fig. 8.2 Inferred collinearity of *COL* genes among genomes of B. oleracea, closely related *Brassica* species, and outgroups. A gene is shown by a pointed rectangle. Homologous genes between neighboring chromosomal regions are linked with lines. At: *Arabidopsis thaliana*; BnA: *B. napus* subgenome A; BnB: *B. napus* subgenome B; Br: *B. rapa*; Bo: *B. oleracea*; Cp: *C. pepo*

is well supported by gene-collinearity-linked orthology and paralogy. We found that after the Brassicapaleotriplication, 57.6% of triplicated gene copies have been preserved, meaning 42.4% copies lost, whereas the gene copy retention and loss rates in *B. oleracea* are 42.4% and 57.6%, respectively. This suggests that *B. oleracea* has lost more copies of *COL* genes than *B. rapa*, whereas allotetraploid *B. napus* does not show a net loss of *COL* genes after allopolyploidization.

8.3.5 Inference of Adaptive Selection

In summary, we found considerable level of *COL* gene losses in *Brassica oleracea* and its closest relatives. This may be caused by the instability of genomes after whole-genome duplication/triplication. In the meantime, natural selection may work to remove many redundant genes. Multiple gene copies existing in the same genome may increase instability of the genome, although they also provide more opportunities

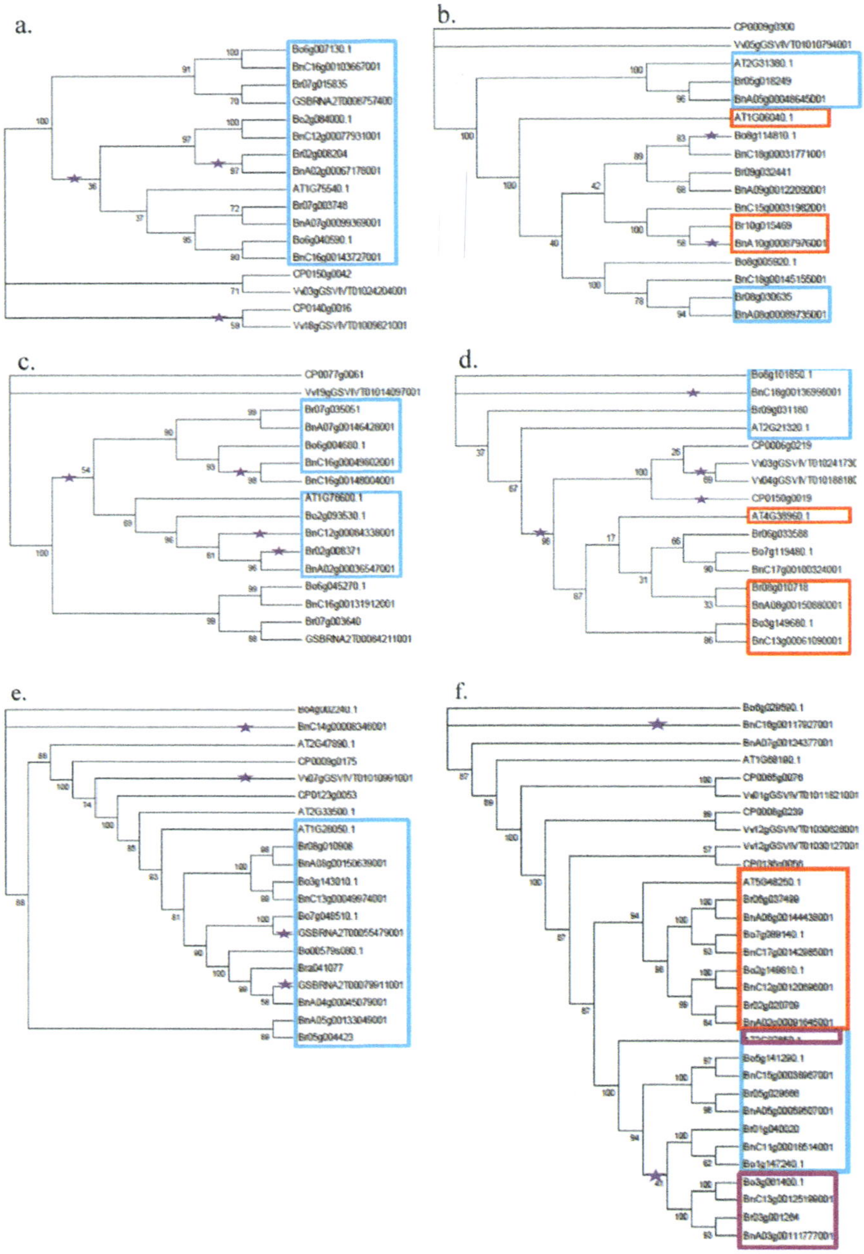

Fig. 8.3 Phylogeny, collinearity, and selection analysis of nine groups of *COL* genes in *B. oleracea* compared to *B. rapa*, *B. napus*, *A. thaliana*, *V. vinifera*, and *C. pepo*. Genes circled in the same colored rectangles share collinearity among genomes or within a genome. Stars on branches show positively selected genes during evolution

for genetic mutation and neofunctionalization. Nevertheless, the loss of some gene copies possibly balances genomic stability and genetic innovation.

To investigate whether and when natural selection had acted on the evolution of the *COL* gene family, we performed selection pressure analyses within each group by using the PAML

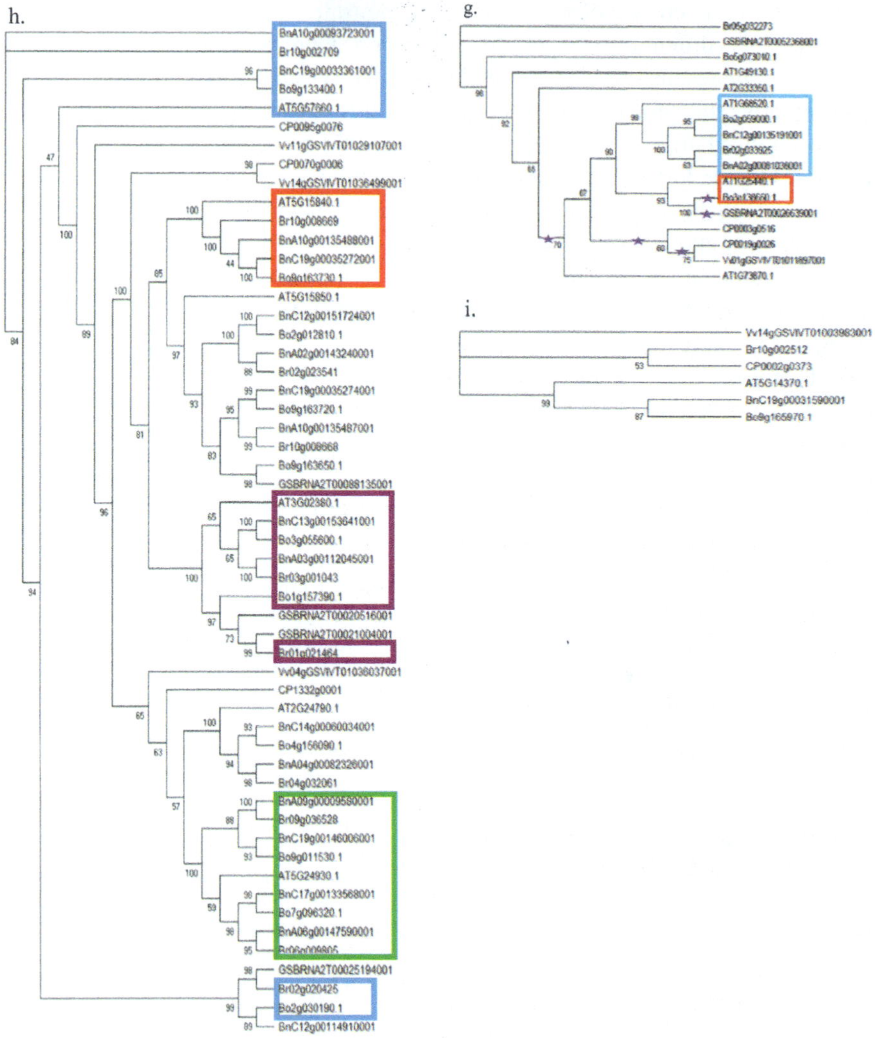

Fig. 8.3 (continued)

program (Fig. 8.3). A few positive selected genes and ancestral nodes were inferred along the phylogenetic tree. For example, a likelihood test showed that in group1 the *B. napus COL* gene BnA10g00087976001 exerted positive selection pressure, while its ortholog in *B. rapa* (Br10g015469) was not positively selected. Similarly, we found *B. oleracea* gene Bo8g114810.1 was positively selected but its ortholog in *B. napus* (BnC18g00031771001) was not. This indicates that polyploidy-induced homologs might be alternatively positively selected in different plants.

8.4 Discussion and Conclusions

8.4.1 Genome Redundancy and Genome Stability

Copy number changes, phylogeny, and evolution of *COL* genes are reported to be related to plant flowering (Fornara et al. 2010). We found extensive gene loss of this large gene family in the paleopolyploid genomes of *B. oleracea* and its closest relatives. This shows the plasticity of plant genomes at the gene level. The Brassica diploid species *B. rapa* and *B. oleracea* have preserved appreciably different rates of genes (57.6% vs. 42.4%), suggesting that *B. rapa* might be more conservative in genome composition. Genes may be removed from a genome due to two possibilities. On one hand, high redundancy of genes may have to be lowered in order to regain genome stability (Freeling et al. 2012). On the other hand, redundant genes could be evolutionarily more vulnerable because they are likely to accumulate more mutations (McGrath et al. 2014). Therefore, it is not unexpected to find appreciable numbers of *COL* gene losses after polyploidy. However, we found the main gene losses occurred after the ancestral Brassica triplication, but not after the formation of the allotetraploid, *B. napus*, which merged the ancestral *B. rapa* and *B. oleracea* genomes together. Therefore, we figure that it is genome instability caused gene losses but not evolutionary selection, which would be expected to drive gene evolution, but is not that effective as compared to genome instability.

8.4.2 Homoeologous Recombination and Genome Stability

Genome instability can be anticipated after each round of polyploidy and is often linked to homo(eo)logous recombination between duplicated copies of chromosomes (Gaeta and Chris 2009). It was reported that artificial wheat could lose 15% of DNA in F_2 plants, and homoeologous recombination was used to explain the instability (Feldman et al. 2012). As a pair of fossil chromosomes, *Oryza sativa* (rice) chromosomes 11 and 12 result from duplication ~ 100 million years ago by tetraploidization in a common grass ancestor, and gene loss together with gene conversion has subsequently occurred in *Oryza* species and possibly other grasses (Wang et al. 2011a, 2015a). A comparison to sorghum indicated that 45% of genes have been removed from one pair of terminal regions of the two rice chromosomes (Wang et al. 2011a). The two homoeologous chromosomes provided a chance to give a glance at the instability of the ancestral genome shortly after the whole-genome duplication. Comparatively, the *B. napus* genome is relatively stable, possibly due to its relatively highly divergent subgenomes, restricting the level of homoeologous recombination, although, which does exist and result in cases of converted genes (Chalhoub et al. 2014).

References

Chalhoub B, Denoeud F, Liu S, Parkin IA, Tang H, Wang X, Chiquet J, Belcram H, Tong C, Samans B et al (2014) Plant genetics. Early allopolyploid evolution in the post-Neolithic *Brassica napus* oilseed genome. Science 345:950–953

Cheng F, Liu S, Wu J, Fang L, Sun S, Liu B, Li P, Hua W, Wang X (2011) BRAD, the genetics and genomics database for Brassica plants. BMC Plant Biol 11:136

Edgar RC (2004) MUSCLE: a multiple sequence alignment method with reduced time and space complexity. BMC Bioinform 5:113

Feldman M, Levy AA, Fahima T, Korol A (2012) Genomic asymmetry in allopolyploid plants: wheat as a model. J Exp Bot 63:5045–5059

Fornara F, de Montaigu A, Coupland G (2010). SnapShot: control of flowering in Arabidopsis. Cell 141 (550):550e 551–552

Freeling M, Woodhouse MR, Subramaniam S, Turco G, Lisch D, Schnable JC (2012) Fractionation mutagenesis and similar consequences of mechanisms removing dispensable or less-expressed DNA in plants. Curr Opin Plant Biol 15:9

Gaeta RT, Chris Pires J (2009) Homoeologous recombination in allopolyploids: the polyploid ratchet. New Phytol 186:18–28

Goodstein DM, Shu S, Howson R, Neupane R, Hayes RD, Fazo J, Mitros T, Dirks W, Hellsten U, Putnam N et al (2012) Phytozome: a comparative platform for green plant genomics. Nucleic Acids Res 40:D1178–D1186

Guindon S, Dufayard JF, Lefort V, Anisimova M, Hordijk W, Gascuel O (2010) New algorithms and methods to estimate maximum-likelihood phylogenies: assessing the performance of PhyML 3.0. Syst Biol 59:307–321

Jiao Y, Leebens-Mack J, Ayyampalayam S, Bowers JE, McKain MR, McNeal J, Rolf M, Ruzicka DR, Wafula E, Wickett NJ et al (2012) A genome triplication associated with early diversification of the core eudicots. Genome Biol 13:R3

Lee TH, Tang H, Wang X, Paterson AH (2013) PGDD: a database of gene and genome duplication in plants. NucleicAcids Res 41:D1152–D1158

Li Q, Zhang N, Zhang L, Ma H (2015) Differential evolution of members of the rhomboid gene family with conservative and divergent patterns. New Phytol 206:368–380

Liu S, Liu Y, Yang X, Tong C, Edwards D, Parkin IA, Zhao M, Ma J, Yu J, Huang S et al (2014) The Brassica oleracea genome reveals the asymmetrical evolution of polyploid genomes. Nat Commun 5:3930

McGrath CL, Gout JF, Johri P, Doak TG, Lynch M (2014) Differential retention and divergent resolution of duplicate genes following whole-genome duplication. Genome Res 24:1665–1675

Mun JH, Yu HJ, Park S, Park BS (2009) Genome-wide identification of NBS-encoding resistance genes in Brassica rapa. Mol Genet Genom. https://doi.org/10.1007/s00438-009-0492-0

Olaitan AO, Diene SM, Assous MV, Rolain JM (2015) Genomic Plasticity of Multidrug-resistant NDM-1 Positive Clinical Isolate of Providencia rettgeri. Genome Biol Evol. https://doi.org/10.1093/gbe/evv195

Proost S, Van Bel M, Sterck L, Billiau K, Van Parys T, Van de Peer Y, Vandepoele K (2009) PLAZA: a comparative genomics resource to study gene and genome evolution in plants. Plant cell 21:3718–3731

Putterill J, Laurie R, Macknight R (2004) It's time to flower: the genetic control of flowering time. BioEssays 26:363–373

Ratnaparkhe MB, Wang X, Li J, Compton RO, Rainville LK, Lemke C, Kim C, Tang H, Paterson AH (2011) Comparative analysis of peanut NBS-LRR gene clusters suggests evolutionary innovation among duplicated domains and erosion of gene microsynteny. New Phytol 192:164–178

Song X, Duan W, Huang Z, Liu G, Wu P, Liu T, Li Y, Hou X (2015) Comprehensive analysis of the flowering genes in Chinese cabbage and examination of evolutionary pattern of *CO*-like genes in plant kingdom. Sci Rep 5:14631

Tamura K, Stecher G, Peterson D, Filipski A, Kumar S (2013) MEGA6: molecular evolutionary genetics analysis version 6.0. Mol Biol Evol 30:2725–2729

Tang H, Bowers JE, Wang X, Ming R, Alam M, Paterson AH (2008) Synteny and collinearity in plant genomes. Science 320:486–488

Velasco R, Zharkikh A, Affourtit J, Dhingra A, Cestaro A, Kalyanaraman A, Fontana P, Bhatnagar SK, Troggio M, Pruss D et al (2010) The genome of the domesticated apple (*Malus* x *domestica* Borkh.). Nat Genet 42:833–839

Wang JP, Yu JG, Li J, Sun PC, Wang L, Yuan JQ, Meng FB, Sun SR, Li YX, Lei TY et al (2018) Two likely auto-tetraploidization events shaped kiwifruit genome and contributed to establishment of the Actinidiaceae family. Science **7**:230–240

Wang X, Guo H, Wang J, Lei T, Liu T, Wang Z, Li Y, Lee TH, Li J, Tang H et al (2016a) Comparative genomic de-convolution of the cotton genome revealed a decaploid ancestor and widespread chromosomal fractionation. New Phytol 209:1252–1263

Wang X, Tang H, Paterson AH (2011a) Seventy million years of concerted evolution of a homoeologous chromosome pair, in parallel, in major Poaceae lineages. Plant Cell 23:27–37

Wang X, Wang H, Wang J, Sun R, Wu J, Liu S, Bai Y, Mun JH, Bancroft I, Cheng F et al (2011b) The genome of the mesopolyploid crop species *Brassica rapa*. Nat Genet 43:1035–1039

Wang X, Wang J, Guo H, Lee T, Liu T, Jin D, Paterson AH (2015a) Genome alignment spanning major poaceae lineages reveals heterogeneous evolutionary rates and alters inferred dates for key evolutionary events. Mol Plant 8:14

Wang X, Wang Z, Guo H, Zhang L, Wang L, Li J, Jin D, Paterson AH (2016b) Telomere-centric genome repatterning determines recurring chromosome number reductions during the evolution of eukaryotes. New Phytol 205:12

Wang Y, Wang X, Lee TH, Mansoor S, Paterson AH (2013) Gene body methylation shows distinct patterns associated with different gene origins and duplication modes and has a heterogeneous relationship with gene expression in *Oryza sativa* (rice). New Phytol 198:274–283

Wang Z, Zhou Z, Liu Y, Liu T, Li Q, Ji Y, Li C, Fang C, Wang M, Wu M et al (2015b) Functional evolution of phosphatidylethanolamine binding proteins in soybean and Arabidopsis. Plant Cell 27:323–336

Yang S, Gu T, Pan C, Feng Z, Ding J, Hang Y, Chen JQ, Tian D (2008) Genetic variation of NBS-LRR class resistance genes in rice lines. Theor Appl Genet 116:165–177

Structural and Functional Genomic Resources and Database Development

9

Jingyin Yu

Abstract

In-depth analysis of the *Brassica oleracea* genome is continually improving the structural and functional genomic resources and databases available to the *Brassica oleracea* research community. In this chapter, we highlight Bolbase as one such example. Bolbase is a comprehensive database which provides access to *B. oleracea* genomic and comparative genomic information. The *B. oleracea* genomic data includes nine fully assembled chromosomes and 1,848 scaffolds, with 45,758 protein-coding genes, 13,382 transposable elements, and 3,581 non-coding RNAs. Comparative genomic information includes syntenic regions among *B. oleracea*, *B. rapa,* and *A. thaliana*, synonymous (Ks) and non-synonymous (Ka) substitution rates between orthologous gene pairs, gene families, or clusters, and differences in quantity, category, and distribution of transposable elements on different chromosomes. Additionally, Bolbase also provides search and data mining tools, including a keyword search, a local BLAST server, and a customized GBrowse tool, which can extract annotations of genomic components identify similar sequences, and visualize syntenic regions among different plant species.

J. Yu (✉)
Key Laboratory of Biology and Genetic Improvement of Oil Crops, The Ministry of Agriculture and Rural Affairs; Oil Crops Research Institute, Chinese Academy of Agricultural Sciences, Wuhan, PR 430062, China
e-mail: yujingyin@caas.cn; jy526@cornell.edu

Boyce Thompson Institute for Plant Research, Cornell University, Ithaca, NY 14853, USA

9.1 Introduction

Over the past decades, many freely available resources for *Brassica* genomes have been built with improved access which further aids *Brassica* research. This includes general integrated databases, such as The National Center for Biotechnology Information (NCBI) (http://www.ncbi.nlm.nih.gov/) (Sayers et al. 2011), The European Bioinformatics Institute, part of the European Molecular Biology Laboratory (EMBL-EBI) (http://www.ebi.ac.uk/) (Abbott 1999) and The DNA Data Bank of Japan (DDBJ) (http://www.ddbj.nig.ac.jp/) (Mashima et al. 2016), as well as Brassica-specific databases, including the *Brassica oleracea* genome database (Bolbase) (http://ocri-genomics.org/bolbase) (Yu et al. 2013), the Brassica transposable elements database (BrassicaTED) (http://im-crop.snu.ac.kr/BrassicaTED) (Murukarthick et al. 2014), Brassica.info (http://www.brassica.info), The Brassica Genome Gateway (http://brassica.nbi.ac.uk/), and The Brassica database (BRAD) (http://brassicadb.org/brad/) (Cheng et al. 2011). Brassica.info

S. Liu et al. (eds.), *The Brassica oleracea Genome*, Compendium of Plant Genomes,
https://doi.org/10.1007/978-3-030-31005-9_9

(http://www.brassica.info) is an integrated database of the genus Brassica which collates and exchanges open-source information relating to Brassica genomics and genetics, on behalf of the Multinational Brassica Genome Project (MBGP). This database contains registries of reference datasets and nomenclature standards, as well as searchable access to relevant databases including a comprehensive compilation of ongoing public domain genome sequencing and a few available publications. The *Brassica* Genome Gateway is an integrated database collating the bacterial artificial chromosome (BAC), single nucleotide polymorphism (SNP) and 95 k unigene sets for the genus *Brassica*, as well as providing news and updates concerning the BBSRC *Brassica* IGF Project and Multinational *Brassica* Genome Project (MBGP). These open-source databases of *Brassica oleracea* genomics brought great convenience for scientists in this community.

In this chapter, we will introduce the use of Bolbase as an example for providing insights into the structural and functional genomics of *Brassica oleracea*. Bolbase is a comprehensive, user-friendly bioinformatics database which allows access to, as well as searching and visualization of *B. oleracea* genomic sequence data and comparative genomic information. The availability and access to this data can improve the current understanding of functional annotation and the evolutionary history of the *B. oleracea* genome. Bolbase will serve as an infrastructure to study gene function, comparative genome, and evolutionary relationship in closely related Brassicaceae species, as well as improve advances for the study of *Brassica* molecular breeding.

9.2 Bolbase Structure

Bolbase built four main functional sections including brief introduction (home unit), main information browsing (browse and synteny units), main tools (local search and customized third party software), and assistance information (document and help units) (Fig. 9.1). For main information browsing, Bolbase provides different *B. oleracea* phenotypes, genome information, pathway annotation, genome visualization, as well as syntenic regions among *B. oleracea*, *B. rapa,* and *A. thaliana* with the information of synonymous (Ks) and non-synonymous (Ka) substitution rates between orthologous gene pairs, gene families or clusters, and differences in quantity, category, and distribution of transposable elements on different chromosomes. Using local searchable tools, users can extract annotations of genome components, identify similar sequences, and visualize syntenic regions among different plant species. Assistance information provides data statistics of *B. oleracea* genomic data and comparative genomics, publication and resource links and provides access to the Bolbase manual, feedback, and downloadable data (Yu et al. 2013).

9.3 Main Features

Bolbase has collated the complete *B. oleracea* assembled genome sequences, including nine pseudo-molecular chromosomes, 1,848 scaffolds, with all genome components, comprising 45,758 putative genes, 13,382 transposable elements, and 3,581 non-coding RNAs. Each annotated genomic component was supplied extensive annotation and external crosslinks to public resource databases. Besides that, Bolbase provides a powerful synteny analysis among *B. oleracea*, *B. rapa*, and *A. thaliana* genomes from BRAD (http://brassicadb.org/brad/, v1.0) (Wang et al. 2015) and TAIR (http://www.arabidopsis.org, TAIR9) (Lamesch et al. 2012), respectively (Yu et al. 2013).

9.3.1 Components

Bolbase collected a total of 45,758 putative genes with detailed functional annotations. Putative genes with a variety of architectonic types, such as gene families, orthologous groups, tandem arrays, and corresponding coordinates on pseudo-molecular chromosomes or scaffolds are freely available for the community. Each putative

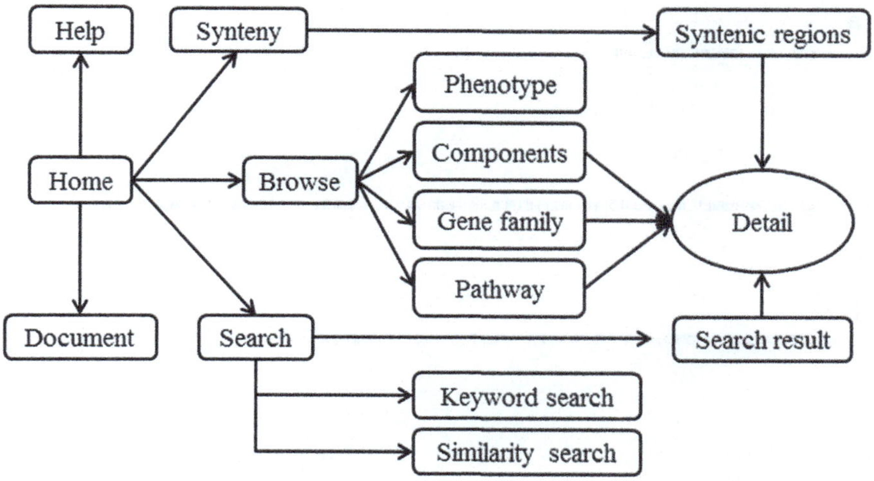

Fig. 9.1 Schematic illustration of Bobase sitemap

gene was annotated using public databases or web services to obtain a comprehensive functional overview. A total of 13,382 transposable elements in *B. oleracea* genome were deposited with two major classes: retrotransposons (Class I transposons) and DNA transposons (Class II transposons). Additional categories, such as long terminal repeat retrotransposons (LTR-RTs), short interspersed nuclear elements (SINEs), long interspersed nuclear elements (LINEs), hAT, Mutator, Tc1-Mariner, Pong, CACTA, Helitron, PIF-Harbinger, and miniature inverted repeat transposable elements (MITEs) are hierarchically listed. Moreover, extensive information for different superfamilies or families of LTR-RT elements is supplied clearly. Bolbase has compiled 3,581 non-coding RNAs through their sequence similarities and conserved motifs: 312 microRNAs (miRNAs), 1,434 small nuclear RNAs (snRNAs: CD-box, HACA-box, and splicing), 517 ribosomal RNAs (rRNAs: 18S, 28S, 5.8S, and 5S), and 1,318 transfer RNAs (tRNAs) (Yu et al. 2013).

9.3.2 Synteny Analysis

To understand genome structure and syntenic relationship in different species, syntenic regions among *Brassica* species and *A. thaliana* were classified using the MCscanX software and manual curation, and they can be visualized in Wang et al. (2012). Orthologous gene pairs among *B. rapa, B. oleracea, and A. thaliana* genomes were first identified based on an all-against-all BLAST search with an E-value cutoff $\leq$1e–10 between different species from best-reciprocal BLAST hits (Altschul et al. 1997). Then, MCscanX was used to identify syntenic regions among *B. rapa*, *B. oleracea,* and *A. thaliana* genomes, using the parameters e = 1e–20, u = 1, and s = 5, which required a minimum of 5 consecutive orthologous gene pairs in the collinear regions. Finally, 558 syntenic genomic regions, including 22,413 gene pairs, were identified between *B. oleracea* and *A. thaliana*, and 1,034 syntenic genomic regions containing 24,422 gene pairs were defined between *B. oleracea* and *B. rapa* (Fig. 9.2) (Yu et al. 2013).

9.4 Visualization

Bolbase built friendly visualized interfaces, which help users to understand and interpret *B. oleracea* genomics and comparative genomics between *B. oleracea* and its relatives.

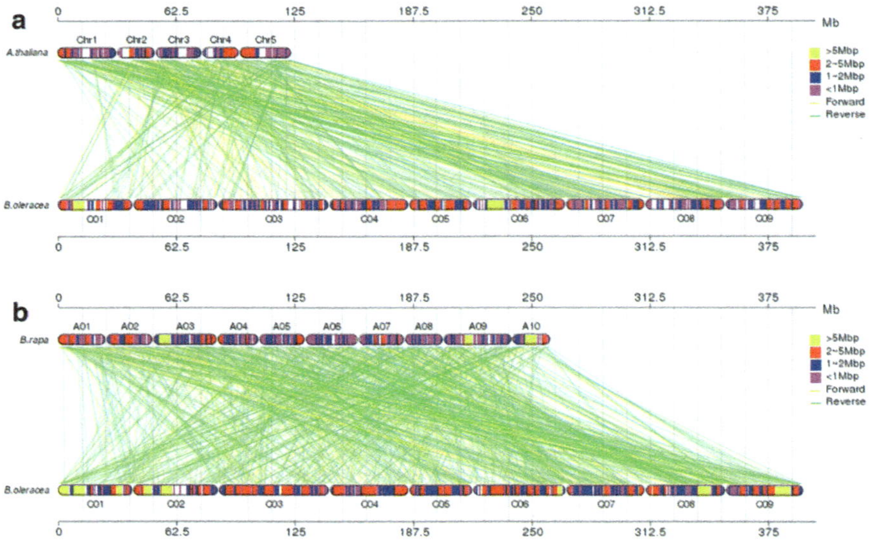

Fig. 9.2 Syntenic regions of *Brassica oleracea* with *Arabidopsis thaliana* and *Brassica rape* genomes. **a** Syntenic regions between *Arabidopsis thaliana* and *Brassica oleracea* genomes. **b** Syntenic regions between *Brassica rapa* and *Brassica oleracea* genomes

9.4.1 Searching

The frontend web pages of genome components are organized by different genome components and displayed by different functional tabs in Bolbase. Each main navigation tab focuses on a certain genome component to allow users to get information from Bolbase. The tab of putative genes is organized by gene families, orthologous groups, tandem arrays, and gene coordinates on pseudo-molecular chromosomes or scaffolds. Repeat element and non-coding RNA tabs are organized by types, categories, or superfamilies. Particularly, Bolbase provides extensive functional annotations for every putative gene that can be divided into four units, including basic information, sequence features, gene clusters containing orthologous groups and tandemly duplicated arrays, and syntenic analyses including orthologs in *B. rapa* and *A. thaliana*, as well as corresponding syntenic regions and triplicated blocks between *A. thaliana* and *Brassica* species. Basic information consists of gene identifier, coordinates, gene model structure (intron/exon boundary, number, length, etc.), and coding nucleotide and protein/peptide sequences. The unit of protein sequence features displays conserved protein domains or motifs predicted by InterProScan (Jones et al. 2014). Additionally, putative genes were annotated and compared with different open-source databases, including Gene Ontology (GO) (Ashburner et al. 2000), Swiss-Prot/TrEMBL (O'Donovan et al. 2002), and Kyoto Encyclopedia of Genes and Genomes (KEGG) (Kanehis and Goto 2000) (Yu et al. 2013).

9.4.2 Browsing

To visualize the syntenic relationship among different species, the syntenic regions in *B. oleracea*, *B. rapa*, and *A. thaliana* are visualized with cartoon chromosomal pictures produced by Perl/CGI, and statistical analyses of gene pairs between species are graphically represented by scatter plots. The syntenic regions between target chromosomes and those of other species will appear after selecting interested chromosomes, revealing gene pairs in each region and their *Ka*, *Ks*, and *Ka*/*Ks* values (Yu et al. 2013).

9.5 Multi-functional Search Tools

Bolbase provides a user-friendly frontend interface to facilitate the retrieval of desired information from *B. oleracea* genome. In this part, utilizing the Bolbase designed searching function units for data mining is discussed. These searching function units include keyword search, orthologous genes search, syntenic regions search, and sequence similarity search.

9.5.1 Keyword Search

The keyword search is a powerful search function to retrieve useful genomic information, such as all types of gene sequences, functional annotations, and homologous genes among different species. This section mainly includes putative gene, transposable element, and non-coding RNA search panel. Putative gene searching will provide users with extensive functional annotations, orthologous genes, and/or tandem arrays, if they exist. By supplying a GO term, an InterPro or a KEGG pathway entry, users can retrieve a group of putative genes in the *B. oleracea* genome. Different types, categories, and superfamilies of transposable elements can be screened in the transposable element search panel. The non-coding RNA search panel is designed to compile information on these genetic elements. The different types or categories of non-coding RNA can be also searched on this panel (Yu et al. 2013).

9.5.2 Orthologous Gene Search

Through comparative analyses among different species, users can further know the genomes of *B. oleracea* and its relatives. Orthologous genes within conserved syntenic regions can be displayed by using localize GBrowse_syn software and supplying a gene name (McKay et al. 2010). This functional unit is contained in the 'Search' section on the main navigation bar. Here, the *B. oleracea* gene Bol007288 was used as an example to show orthologous genes in its relative species. By searching with Bol007288 as a query on the orthologous genes search panel, two orthologs in *A. thaliana* (AT5G06860 and AT3G12090) and two in *B. rapa* (Bra038699 and Bra000594) are retrieved from Bolbase (Yu et al. 2013).

9.5.3 Syntenic Region Search

The customized GBrowse software is included in Bolbase to search syntenic genomic regions or sequences. Through choosing a chromosome from one species, the syntenic genomic regions or sequences within corresponding species are displayed alongside their chromosomal positions. When the hyperlink for the target region is chosen, the syntenic regions in corresponding species will be displayed in this panel (Yu et al. 2013).

9.5.4 Sequence Similarity Search

As sequence and structure vary, the similarity page using BLAST searching can satisfy users with different interests. By supplying a nucleic acid or an amino acid sequence via directly pasting or file uploading to search against the available databases, users can retrieve homologous genes or genomic sequences. This allows for quick comparisons and annotations of query sequences with the deposited data in Bolbase. BLAST hits are returned with hyperlinks to the genes to enable users to quickly acquire their annotations in different genomes (Yu et al. 2013).

9.6 Conclusion and Perspective

Bolbase is a comprehensive database for *B. oleracea* genomics and comparative genomics to facilitate the development of molecular biology and genetics breeding for the genus *Brassica*. Bolbase provides genome-wide datasets, accurate functional annotations, and syntenic analyses for the *B. oleracea* genome and closely related *Brassica* species, which are not available in currently existing open-source databases of Brassica species. The application of Bolbase to *Brassica* study will open new avenues for scientists to explore and understand the morphological diversity and evolution of genome structure after polyploidization. Bolbase will supply a comprehensive starting point for plant molecular biologists and breeders to explore the Brassica genus further, as well as the entire crucifer lineage.

References

Abbott A (1999) EMBL rescue package keeps bioinformatics centre running. Eur Mol Biol Laborat Nature 402(6761):450

Altschul SF, Madden TL, Schaffer AA, Zhang J, Zhang Z et al (1997) Gapped BLAST and PSI-BLAST: a new generation of protein database search programs. Nucleic Acids Res 25(17):3389–3402

Ashburner M, Ball CA, Blake JA, Botstein D, Butler H et al (2000) Gene ontology: tool for the unification of biology. Gene Ontology Consort Nat Genet 25(1):25–29

Cheng F, Liu S, Wu J, Fang L, Sun S et al (2011) BRAD, the genetics and genomics database for Brassica plants. BMC Plant Biol 11:136

Jones P, Binns D, Chang HY, Fraser M, Li W et al (2014) InterProScan 5: genome-scale protein function classification. Bioinformatics 30(9):1236–1240

Kanehis M, Goto S (2000) KEGG: kyoto encyclopedia of genes and genomes. Nucleic Acids Res 28(1):27–30

Lamesch P, Berardini TZ, Li D, Swarbreck D, Wilks C et al (2012) The Arabidopsis Information Resource (TAIR): improved gene annotation and new tools. Nucleic Acids Res 40(Database issue): D1202–D1210

Mashima J, Kodama Y, Kosuge T, Fujisawa T, Katayama T et al (2016) DNA data bank of Japan (DDBJ) progress report. Nucleic Acids Res 44(D1): D51–D57

McKay SJ, Vergara IA, Stajich JE (2010) Using the Generic Synteny Browser (GBrowse_syn). Curr Protoc Bioinformatics. 2010 Sep, Chapter: Unit–9.12

Murukarthick J, Sampath P, Lee SC, Choi BS, Senthi lN, et al (2014) BrassicaTED−a public database for utilization of miniature transposable elements in Brassica species. BMC Res Notes 7:379

O'Donovan C, Martin MJ, Gattiker A, Gasteiger E, Bairoch A et al (2002) High-quality protein knowledge resource: SWISS-PROT and TrEMBL. Brief Bioinform 3(3):275–284

Sayers EW, Barrett T, Benson DA, Bolton E, Bryant SH et al (2011) Database resources of the National Center for Biotechnology Information. Nucleic Acids Res 39 (Database issue): D38–D51

Wang X, Wu J, Liang J, Cheng F (2015) Brassica database (BRAD) version 2.0: integrating and mining Brassicaceae species genomic resources. Database (Oxford), bav039

Wang Y, Tang H, Debarry JD, Tan X, Li J et al (2012) MCScanX: a toolkit for detection and evolutionary analysis of gene synteny and collinearity. Nucleic Acids Res 40(7):e49

Yu J, Zhao M, Wang X, Tong C, Huang S et al (2013) Bolbase: a comprehensive genomics database for Brassica oleracea. BMC Genom 14:664

Impact of the *Brassica oleracea* Genome on Breeding and Crop Improvement

10

J. Fidler, N. Żyła, and D. Babula-Skowrońska

Abstract

Brassica oleracea is one of the world's most valuable vegetable species, making an important contribution to the global human diet. The species has extremely broad genetic and morphological variation along with high diversity for nutritionally interesting compounds. Among them, we can distinguish a high content of protein, vitamins, carotenoids, flavonoids, polyphenols, minerals, and glucosinolates. The latter are unique phytochemicals which, besides their value for human nutritional/health, are also involved in plant defense against fungal and bacterial pathogens. To study the genetic background and to improve these agronomic traits in *B. oleracea*, numerous breeding programs have been performed. The recognition of genetic resources controlling morphological traits responsible for plant architecture, physiological factors such as mineral contents and resistance to diseases facilitate production of new varieties based on gene transfer by backcrossing. The identification of useful diversity for breeding, genetic analysis, marker development, and marker-assisted selection for important traits can profit considerably from the availability of the *B. oleracea* reference genome sequence and other related genomics resources (mapping population, TILLING population, clone libraries). However, a challenge for these programs is the structural complexity of the *B. oleracea* genome and limited gene pool as well as genetic incompatibility within *B. oleracea* genotypes that results from the breeding procedures used to generate them. In this chapter, we summarize the progress in breeding programs aspiring to improving *B. oleracea* varieties. They include morphological traits, minerals as well as resistance to insects and diseases. We discuss that classical genetic analyses aided by modern system biology approaches can advance works on the choice of starting materials to the introduction of new genotypes.

10.1 Introduction

The globally important vegetable species *B. oleracea* has a very wide morphological diversity and its highly diverse vegetable forms make an important contribution to the human diet. Secondary metabolites found in *B. oleracea* leaves and other plant organs have been found to have direct health benefits, including the potential to reduce cancer risk and to alleviate other diseases, such as diabetes and cardiovascular conditions (Bhandari and Kwak 2015; Pagliaro

J. Fidler · N. Żyła · D. Babula-Skowrońska (✉)
Institute of Plant Genetics, Polish Academy of Sciences, Strzeszyńska 34, 60-479 Poznań, Poland
e-mail: dbab@igr.poznan.pl

S. Liu et al. (eds.), *The Brassica oleracea Genome*, Compendium of Plant Genomes,
https://doi.org/10.1007/978-3-030-31005-9_10

et al. 2015; Shah et al. 2016). Brassicas are characterized by a high content of protein, vitamins, carotenoids, flavonoids, polyphenols, minerals, and glucosinolates (GSLs). The latter are unique phytochemicals which, besides their value for human nutritional/health, are also involved in plant defense against fungal and bacterial pathogens (Halkier and Gershenzon 2006; Kopsell and Kopsell 2006).

To study the genetic background and thereby to improve agronomic traits in *B. oleracea*, many short- and long-term breeding programs have been performed. A major challenge for these programs is the structural complexity of the *B. oleracea* genome, which complicates map-based cloning and the identification of genes underlying quantitative trait loci. Another problem is the limited gene pool and genetic incompatibility within *B. oleracea* genotypes that result from the breeding procedures used to generate them. To increase the size of the gene pool, introgression of desirable genes from wild relatives of *B. oleracea* cultivars has been carried out. These wild species provide a source of cytoplasmic and nuclear male sterility, resistance to disease and insects, and intermediate C3-C4 photosynthetic activity, as well as tolerance of cold, salt, and drought conditions. Recently, the *Brassica* C pangenome has been constructed using sequencing data from nine cultivated *B. oleracea* varieties and a wild relative, *B. macrocarpa* (Golicz et al. 2016). Most unique genes relating to disease resistance, flowering time, GSL metabolism, and vitamin biosynthesis have only been found in the wild species, which suggests that these valuable traits have probably been lost during domestication. The identification of gene presence/absence variation (PAV) in this pangenome will be useful for breeding programs aimed at the analysis of heterosis in F_1 hybrids and the identification of complex agronomic traits using association methods (Golicz et al. 2016).

A major goal of *B. oleracea* breeding programs is the production and cultivation of disease-resistant varieties that allow more efficient, economical, and environmentally friendly farming practices. In such programs, the selection step is key, and aims to modify important agronomical traits. The traditional method of selection is based on phenotype, which is both time-consuming and costly. It also suffers from the major disadvantage that it cannot differentiate between hetero- and homozygous plants; it is also difficult to assess the impact of environment on the phenotype. In recent years, new methods of selection have been developed that are effective, cost-effective, co-dominant, and unlimited by environment and stage of plant development. These methods use modern technologies based on next-generation sequencing (NGS) and analysis of whole-genome sequences. They include marker-assisted selection (MAS; Chen et al. 2013; Lv et al. 2013), divergent mass selection (DMS; Sotelo et al. 2016), genome-wide association (GWAS; Thorwarth et al. 2018), and genotyping-by-sequencing (GBS; Stansell et al. 2018).

10.2 Agronomic Traits and Physiological Features

To understand and exploit the genetic control of morphological diversity within *B. oleracea*, collections of germplasms and biparental populations derived from crosses between different morphotypes have been developed (Bothmer et al. 1995; Lazaro and Aguinagalde 1998; Divaret et al. 1999; Sebastian et al. 2000; Smith and King 2000). These projects are supported by NGS-based approaches and most genetic resources derived from the research, such as genetic maps, mapping populations, molecular markers, and also information about quantitative traits and their location in the genome have been deposited in the public databases CropStoreDB (http://www.cropstoredb.org/) and Brassica.info (https://www.brassica.info/).

Most important agronomic traits such as plant height, yield, and maturation period are quantitative traits. Quantitative trait loci (QTLs) are mapped to linkage groups using various mapping populations and types of polymorphic molecular markers such as restriction fragment length polymorphism (RFLP), random-amplified polymorphic DNA (RAPD), and amplified fragment length polymorphism (AFLP). In *B. oleracea*, many genetic studies have focused on the identification of loci controlling morphological variation, as well as developmental and physiological traits. These include features such as plant size, stem-related and heading traits, fertility and seed germination, flowering time, water absorption and photosynthetic utilization efficiency, regeneration capability of tissue cultures in *Agrobacterium*-mediated transformation, and regeneration capability of protoplasts (Table 10.1). Lan and Paterson (2001) analyzed six traits controlling the size of leaves and stems in three F_2 populations (crosses between a rapid-cycling RCB line and three varieties: Pusa Katki, Cantanese, and Bugh Kana). Comparative analysis of 47 QTLs identified in these populations showed variation in 35 different genetic loci and 28 ancestral genes, and thereby allowed QTL-phenotype associations to be characterized in *B. oleracea*. At the same time, Sebastian et al. (2002) analyzed morphological and developmental traits in F_1 population derived from a cross between two different *B. oleracea* types: an annual cauliflower (line derived from the F_1 'Nedcha', CA25) and a biennial Brussels sprout (line derived from the F_1 'Gower', AC498). They found at least 32 QTLs located in all nine linkage groups of the *B. oleracea* genetic map, and they controlled traits related to leaf, flowering, axillary bud, and stem characters by using the multiple-marker regression approach (Table 10.1). Among them, the researchers located a putative QTL for vernalization on linkage group O8 and this region showed homology with earlier detected QTL for this trait on linkage group O9 in *B. napus* (Ferreira et al. 1995a). More recently, the application of MAS methodology based on whole-genome mapping resulted in 144 QTLs being associated with 24 agronomic features relating to plant-type, leaf, and head traits in cabbage (Lv et al. 2016). The authors determined that most QTLs were clustered within the same genomic region and were significantly correlated with each other. This study demonstrates that MAS methodology is useful in cabbage breeding and facilitates an understanding of the genetic control of these agronomic traits. Subsequently, the GBS method has been used to study morphotype diversity in 85 strains of landrace and improved *B. oleracea* broccoli, cauliflower, and Chinese kale (Stansell et al. 2018). The results revealed a complex population structure and established phylogenetic relationships among *B. oleracea* varieties.

A number of genetic studies have identified loci associated with physiological traits in *Brassica* crops. Flowering time and regulation of flowering are key physiological traits characterized by high variability in brassicas, but are strongly dependent on the growing season, specific geographical region, air temperature, vernalization, and day/night length. Temperature and day length are particularly important factors regulating flowering in the *Brassica* species, which are classified into two main groups, namely biennials, which require prolonged exposure to low temperature to initiate flowering (vernalization type, e.g., cabbage, kohlrabi) and annuals, which do not require such exposure to low temperature (non-vernalization type, e.g., broccoli, cauliflower). Some wild *B. oleracea* forms are also perennial. Different approaches and mapping populations derived from crosses between distinct subspecies have been used to identify markers, QTLs, and genes involved in developmental processes, including flowering time (Table 10.1). These include genotypes differing in vernalization dependence, for example, BRA1398 (*B. oleracea* ssp. *capitata* convar. *botrytis* var. *botrytis* L.; annual) and Kashirka (*B. oleracea* ssp. *capitata*; biennal) (Schiessl et al. 2017), and four genotypes of broccoli (E1, E5, E8, and E9) differing in their requirement for cold periods to form heads (Irwin et al. 2016). Among the mapping populations used to identify QTLs controlling flowering time were: an F_2 family derived from a cabbage and broccoli cross (Kennard et al. 1994); F_3 families derived from different crosses between cabbage and broccoli (Camargo and Osborn 1996); doubled haploid (DH) lines derived from the F_1 of a cross between a Chinese kale A12 and a hybrid of calabrese GD33 (Bohuon et al. 1998; Sebastian et al. 2000; Razi et al. 2008); recombinant

Table 10.1 Summary of markers, genes, and QTLs controlling important agronomic traits used in *B. oleracea* genetics and breeding

Population	Traits	Strategy	Localization of linkage group (LG)/chromosomes[a]	Number of loci	Authors
B. oleracea	Variation in shoot mineral concentration (C, Ca, K, Mg, P, N)	Phylogenetic analysis and a pro rata sampling technique	–	–	Broadley et al. (2004)
Diversity fixed foundation set (DFFS) of 376 *B. oleracea* accessions and 74 modern F_1 hybrid cultivars	Genetic variation in shoot Ca and Mg concentration	Variance components analyses, QTL analysis	C02, C06, C07, C08, and C09	Several QTLs	Broadley et al. (2008)
Diversity fixed foundation set (DFFS) of 376 *B. oleracea* accessions and 74 modern F_1 hybrid cultivars; the AGDH population derived from a cross between A12DHd (*B. oleracea* var. *alboglabra*) and GDDH33 (*B. oleracea* var. *italica*, calabrese cultivar, 'Green Duke')	Genetic variation in shoot Zn concentration	Mineral analysis, QTL analysis	C02, C03, C05, C07, and C09	Several QTLs	Broadley et al. (2010)
F_2 population derived from a cross between an elite broccoli breeding line (360–7) (low ability to regenerate shoots from protoplast) and rapid cycling (RC) line of *B. oleracea* (high ability to regenerate shoots from protoplast)	Plant-regeneration ability of protoplasts	Interval mapping, modified bulked segregant analysis	LGCO2 and CO9	2	Holme et al. (2004)
Different *B. oleracea* subsp. *italica* genotypes (purple sprouting broccoli); DH mapping population derived from a cross between lines E5 and E9 of purple sprouting broccoli	Variation in flowering time and response to overwintering	Analysis of polymorphism at *FLC*, QTL analysis, heterologous complementation experiments	LG C02	8 QTLs, 2 *BoFLC.C2* alleles	Irwin et al. (2016)
F_2 population from a cross between the open-pollinated *B.*	Variation in morphological traits: flower development,	Genetic mapping, single factor	LG C01-C09	22	Kennard et al. (1994)

(continued)

Table 10.1 (continued)

Population	Traits	Strategy	Localization of linkage group (LG)/chromo-somes[a]	Number of loci	Authors
oleracea cabbage cultivar 'Wisconsin Golden Acre' and the hybrid broccoli cultivar 'Packman'	leaf dimensions, heading habit, and internode distance	ANOVA, Multi locus models			
Three F_2 populations derived from crosses between rapid-cycling RCB line and three varieties: Pusa Katki, Cantanese and Bugh Kana	Morphological traits, including lamina length, lamina width, petiole length, stem length, stem width and node number	QTL-phenotype associations	Different chromosomes	47	Lan and Paterson (2001)
F_2 population derived from a cross between cabbage line No. 86–16–5 (resistant to race 2 of *Plasmodiophora brassicae*) and a rapid cycling line CrGC No. 85	Clubroot resistance	Genetic mapping	LG C01 and C06	2 dominant QTLs (*CR2a* and *CR2b*) for resistance to race 2	Landry et al. (1992)
F_2 population derived from a cross between lines C1176 (susceptible to clubroot disease) and C1220 (resistant to clubroot disease)	Clubroot resistance	Genotyping-by-sequencing (GBS) technique	LG C02 and C03	2 single major QTLs for Race 2 and Race 9 of *P. brassicae*	Lee et al. (2006)
Three crosses: broccoli x cauliflower, collard x broccoli, and collard x cauliflower	Inheritance of three major genes involved in synthesis of aliphatic GSLs	Segregating analysis, GSL profiling analysis	–	3	Li et al. (2001)
Cabbage (*B, oleracea* var. *capitata* L.)	Molecular mechanism of vernalization	Cloning of *FLC* genes, structural and functional characterizations	–	2	Lin et al. (2005)
230 DH lines derived from a cross between two cabbage lines: 01–20 (highly susceptible to CFW) and 96–100 (highly resistant to CFW)	Cabbage Fusarium Wilt (CFW), transfer of a CFW-resistance gene*FOC1* from resistant line into susceptible line	Combined methodology of microspore culture and MAS, development of markers closely linked to *FOC1*	C06	1	Liu et al. (2017)
F_1 population derived from a cross between lines 99–77 (highly resistant to CFW) and 99–91 (highly susceptible to CFW)	Fusarium wilt resistance	Genetic mapping, identification of InDel markers closely linked to the resistance gene *FOC*	LG C06	*FOC* gene, 7 InDel markers	Lv et al. (2013)

(continued)

Table 10.1 (continued)

Population	Traits	Strategy	Localization of linkage group (LG)/chromosomes[a]	Number of loci	Authors
F_1 population derived from a cross between lines 99–77 (highly resistant to CFW) and 99–91 (highly susceptible to CFW)	Resistance to Fusarium wilt	Fine mapping, analysis of a candidate gene for *FOC1*	LG C06	4 R-genes, including candidate gene for *FOC1*	Lv et al. (2014a)
DH population 01–20 × 996–100 derived from a cross between lines 01–20 (early-maturing spring cabbage inbred line) and 96–100–308 (early-maturing spring cabbage inbred line)	Heading traits including head maturation period, head weight, core length, head vertical diameter and the ratio of Cl to Hvd	Genetic mapping, QTL analysis	C01, C02, C03, C05, C06, C07 and C09	13	Lv et al. (2014b)
DH population 01–20 × 996–100 derived from a cross between lines 01–20 (early-maturing spring cabbage inbred line) and 96–100–308 (early-maturing spring cabbage inbred line)	Analysis of agronomic traits relating to plant-type, leaf and head	Genetic mapping, QTL analysis	LG C01-C09	144	Lv et al. (2016)
F_2 population derived from a cross between resistant accession of wild *B. oleracea* (*B. incana*) and a susceptible accession of cultivated *B. oleracea* var. *alboglabra*	Sclerotinia stem rot resistance	Phenotypic measurements and genetic mapping	LG C01, C03, C04, C06, C07 and C09	12 QTLs for leaf resistance and 6 QTLs for stem resistance	Mei et al. (2013)
F_3 progenies from a cross between cabbage cultivar 'Anju' (resistant line) and broccoli cultivar 'Green Comet' (GC) (susceptible line)	Clubroot resistance	QTLs, MAS	LGs: C02, C03, C05, C07	5	Nagaoka et al. (2010)
F_2 mapping population derived from a cross of *B. oleracea* cv. Green Comet (broccoli; non-vernalization type) and *B. oleracea* cv. Reiho (cabbage; vernalization type)	Vernalization flowering time	Gene cloning, QTL analysis and genotyping	LG C02, C03, C06, C08 and C09	6 QTLs for flowering time and 4 *FLC* genes	Okazaki et al. (2007)

(continued)

Table 10.1 (continued)

Population	Traits	Strategy	Localization of linkage group (LG)/chromosomes[a]	Number of loci	Authors
188 F_2 lines derived from a cross between cabbage inbred lines "748" (head-splitting-resistant) and "747" (head-splitting-susceptible)	Head-splitting	QTL analysis	LG C02, C04 and C06	6	Pang et al. (2015)
F_2 population derived from a cross between a clubroot-resistant *B. oleracea* inbred line '263' and a susceptible line 'GZ87'	Clubroot resistance	Genotyping with the 60 K *Brassica* SNP array	LG C01, C02, C03, C04, C06, C08	6 CR-associated traits, 23 QTLs	Peng et al. (2018)
F_2 population derived from a cross between the Fusarium wilt-susceptible broccoli cultivar Green Comet and wilt-resistant cabbage cultivar Anju	Fusarium resistance	Segregation test and QTL analysis	LG C07	1	Pu et al. (2012)
79 recombinant backcross substitution lines from a cross between A12DHd (*B. oleracea* var. *alboglabra*) and GDDH33 (*B. oleracea* var. *italica*, calabrese cultivar, 'Green Duke')	Flowering time	QTL mapping	LG C01, C02, C03, C05 and C09	11	Rae et al. (1999)
Two different genotypes of *B. oleracea* ssp. *capitata* BRA1398 (annual) and Kashirka (biennial, a late flowering Siberian kale)	Genetic variation in central flowering time genes	Sequence capture followed by Illumina sequencing	C02, C03, C04, C05, C06, C08 and C09	35 flowering-time regulatory genes	Schiessl et al. (2017)
Mapping population derived from a cross between a cauliflower line 'Nedcha', (CA25), and Brussels sprout line 'Gower' (AC498)	Analysis of morphological and developmental traits: leaf, flowering, axillary bud and stem characters	Multiple-marker regression approach, QTL	LG C01-C09	27	Sebastian et al. (2002)

(continued)

Table 10.1 (continued)

Population	Traits	Strategy	Localization of linkage group (LG)/chromosomes[a]	Number of loci	Authors
Ten Chinese cabbage inbred lines (RJKB-T4, 13–15, 17, 21–25) and commercial cultivar 'Chiifu'; F_2 populations derived from a cross between *Fusarium* yellows susceptible and resistant lines, RJKB-T24 × RJKB-T23 and RJKB-T22 × RJKB-T21	*Fusarium* yellows resistance	Transcriptome of Fusarium yellows resistant and susceptible lines using RNA-seq, genetic mapping	–	2	Shimizu et al. (2014a)
F_2 population derived from resistant cabbage (AnjuP01) and susceptible broccoli (GCP04)	*Fusarium* yellows resistance	Genetic mapping	LG C07	1	Shimizu et al. (2014b)
DH population CA25 x BI87053 derived from a cross between cauliflower variety Nedcha and Calabrese broccoli	Heading morphology, analysis of the CAL/AP1 model for curd-like phe- notype	Genetic linkage analysis	LG C03 and C06	2	Smith and King (2000)
The A12DH mapping population derived from a cross between A12DHd (*B. oleracea* var. *alboglabra*) and GDDH33 (*B. oleracea* var. *italica*)	Genetic control and heritability of *Agrobacterium tumefaciens* susceptibility	Genetic components analysis of *A. tumefaciens* susceptibility in *B. oleracea*, mapping QTL	LG C09	1	Sparrow et al. (2004a)
eight DH genotypes from the A12DH mapping population; four substitution lines −A12DHd genome with substituted regions of the GDDH33 genome	Identification of *B. oleracea* genotypes amenable to transformation	Use of phenotypic tissue culture markers to identify *B. oleracea* genotypes most amenable to transformation	–	–	Sparrow et al. (2004b)
DH population derived from a cross between inbred cabbage lines: 79–156 (head splitting-susceptible) and 96–100 (head splitting-resistant)	Head-splitting resistance (HSR)	Mixed major gene/polygene inheritance analysis and QTL mapping	LG C03, C04, C07 and C09	9	Su et al. (2015)

(continued)

Table 10.1 (continued)

Population	Traits	Strategy	Localization of linkage group (LG)/ chromosomes[a]	Number of loci	Authors
DH mapping population Bi x Gr derived from a cross between DH-line Bi from cabbage landrace Bindsachsenerand DH-line Gr from broccoli cv 'Greenia'	Clubroot resistance	Multiple QTL mapping approach	LG C01, C03 and C06	4	Voorrips et al. (1997)
a diversity fixed foundation set (DFFS) of 376 *B. oleracea* accessions and 74 commercial genotypes, the AGDH population derived from a cross between A12DHd (*B. oleracea* var. *alboglabra*) and GDDH33 (*B. oleracea* var. *italica,* calabrese cultivar, 'Green Duke')	Genetic factors affecting K utilization efficiency	QTL	LG C01, C02, C03, C04, C07 and C09	Several QTLs	White et al. (2010)

[a]The nomenclature of chromosomes and linkage groups has been unified

backcross substitution lines from a cross between GDDH33 (*B. oleracea* var. *italica*) and A12DH (*B. oleracea* var. *alboglabra*) (Rae et al. 1999); a cross between broccoli (*B. oleracea* cv. Green Comet; non-vernalization type) and cabbage (*B. oleracea* cv. Reiho; vernalization type) (Okazaki et al. 2007); and a DH mapping population derived from a cross between E5 and E9 broccoli genotypes (Irwin et al. 2016) (Table 10.1). QTLs relating to flowering control features were described in several reports. Bohuon and coworkers (1998) identified six QTLs for flowering time and mapped them to chromosomes O2, O3, O5, and O9, while Rae and colleagues (1999) found 11 QTLs on chromosomes O1, O2, O3, O5, and O9. Irwin and co-workers (2016) mapped 8 QTLs controlling vernalization response under both glasshouse and field conditions. One of these, a major QTL for heading date variation was mapped to *B. oleracea* chromosome 2 during exposure to different cold conditions. This region includes one of the genes involved in the control of flowering time, i.e. *BoFLC4* (*FLOWERING LOCUS C*). The lack of uniform nomenclature significantly limits the ability to compare data across different studies, however. Therefore, only a small number of the QTLs can be confirmed as identical. This is an issue, for example, in the case of five QTLs that have been mapped to similar regions in two independent works using different lines from the same cross (Bohuon et al. 1998; Rae et al. 1999).

Additionally, one of the five homologs of the *BoFLC* gene located on the *B. oleracea* chromosome 2 has several different names: Lin et al. (2005) named it *BoFLC4*, but it is also known as *BoFLC2* (Ridge et al. 2015) and *BoFLC.C2* (Irwin et al. 2016).

Further studies concern the identification and isolation of genes involved in the control of flowering time in *B. oleracea*. The main candidate genes are *FLC* (*FLOWERING LOCUS C*), encoding a MADS-box transcription factor, *FT* (*FLOWERING LOCUS T*), and *CO* (*CONSTANS*), which occur in multiple copies in the *B. oleracea* genome. However, other genes, such as those involved in sugar signaling and stress-response pathways, may also modulate the flowering response (Blümel et al. 2015). Initially, some *FLC* homologs were identified in the *B. oleracea* genome: *BoFLC1*, *BoFLC3*, *BoFLC4*, and *BoFLC5* (Schranz et al. 2002; Lin et al. 2005). Okazaki and colleagues (2007) also isolated four copies of the *BoFLC* gene. Together, five copies of the *BolFLC* gene on chromosomes C02, C03, and C09 have been identified in *B. oleracea*. Many sequence polymorphisms resulting in premature stop codons, non-synonymous single nucleotide polymorphisms (SNPs), and differential promoter structure have been found in these genes (Okazaki et al. 2007; Irwin et al. 2016). In addition, four copies of *BolFT* on chromosomes C02, C04, and C06, and three copies of *BolCO* on chromosomes C01, C03, and C09, have been reported. Subsequent research has focused on the relationship between chromosomal regions with QTLs and the genes involved in the regulation of flowering process. Okazaki and colleagues (2007) found no association between three genes (*BoFLC1*, *BoFLC3*, and *BoFLC5*) and six QTLs controlling flowering time. However, a fourth gene, *BoFLC2*, was linked to a QTL and genotyping of F_2 plants at the *BoFLC2* locus confirmed the involvement of this gene in the control of flowering time in *B. oleracea*. The *BolFLC2* variation resulted from a SNP in exon 4. A loss-of-function mutation in *BoFLC2* confirmed its participation in earliness in annual cauliflower, reflecting a similar result for *AtFLC* in *A. thaliana* (Ridge et al. 2015). Interestingly, repression of *BoFLC2* function activates the expression of *BoFT* genes. Also, heterologous complementation of *A. thaliana* genotype Columbia *FRI flc2* by one of two alleles of *BoFLC4*, *BoFLC.C2*^{E5} or *BoFLC.C2*^{E9}, confirmed that allelic variation at *BolFLC4* can influence sensitivity to environmental temperature during vernalization and is responsible for heading date variation (Irwin et al. 2016). Studies based on genotype-specific and QTL-based phenology models showed the existence of *FLC*-independent vernalization pathways in *B. oleracea*, as reported in other plant species (Uptmoor et al. 2012). The basis for the distinct expression dynamics of the *BoFLC4* alleles, both cold-induced *BoFLC* repression and subsequent reactivation on return to warmer conditions, is *cis* variation (assigned to a non-coding intronic polymorphism). This polymorphism influences the degree of epigenetic silencing of *BoFLC4* expression in response to low temperature; a similar result was observed earlier in *A. thaliana* (Bastow et al. 2004). As in previous studies, silencing the function of both *BoFLC.C2* alleles caused the activation to different levels of both the *BoFTC2* and *BoFTC6* genes, but only transiently. Variations in SNPs, insertion/deletion (InDels) and copy number of genes known as central flowering regulators, which also represent the total flowering time network, have been analyzed using sequence capture followed by Illumina sequencing in two phenotypes of *B. oleracea* (Schiessl et al. 2017). The selected 35 genes controlled, *inter alia*, the circadian clock, vernalization, and photoperiod pathways, as well as gibberellin signaling. These studies showed that the late-flowering phenotype of Siberian kale Kashirka results from the putative duplication of *Bol.FLC.C09b* or the R24Q and A75D mutations in *Bol.FLC.C03* and *Bol.FLC.C02*, respectively. Such changes in the sequences of key genes can, therefore, affect *B. oleracea* phenotype and illustrate a framework for flowering gene variation. This knowledge about regulation of flowering time and the genes involved is crucial for the improvement of *Brassica* species. For example, *cis* polymorphism in *BoFLC* should be taken into account in the selection of commercial and/or breeding *B. oleracea* accessions, especially broccoli, with an extended season.

Green leafy vegetables such as *B. oleracea* are an important source of minerals and are essential components of the human diet. For example, kale (*B. oleracea* var. *acephala*) contains 130 mg Ca 100 g^{-1} fresh weight, 34 mg Mg 100 g^{-1} f. w. and nitrate at 5758 mg/kg (Food Standards Agency). Broadley and coworkers (2004) analyzed the shoot mineral concentration of minerals such as Ca, K, Mg, Na, P, and N in angiosperms including Brassicales. They found that shoot mineral concentrations differed significantly between subtaxa. In Brassicales, the mineral concentration of Ca, K, and Mg was 2.80% dry weight (d. w.), 4.23% d. w. and 0.45% d. w., respectively. Later, these studies were extended to a set of 376 accessions of *B. oleracea* including wild relatives derived from natural populations (Broadley et al. 2008). These data also showed significant differences in shoot mineral concentrations between subtaxa. The closely related cabbage and Savoy cabbage had the highest shoot Ca and Mg concentrations, as opposed to kohlrabi, which showed the lowest values. Further studies focused on the identification of QTLs associated with shoot Ca, Mg, and Zn concentrations in the integrated AGDH mapping population of *B. oleracea* (Sebastian et al. 2000; Broadley et al. 2008; Broadley et al. 2010). These QTLs have been detected on linkage groups C02, C06, C07, C08, and C09 for Ca and Mg concentrations and on linkage groups C02, C03, C05, C07, and C09 for Zn concentration. The authors also found links between a region including a major QTL on *B. oleracea* chromosome 9 and a 0.41 Mb region on chromosome 5 in *A. thaliana*, which contains two genes encoding the divalent cation-permeable, plasma-membrane proteins, *AtZIP12* and a P-type ATPase. In other studies, White and colleagues (2010) analyzed genetic factors responsible for K utilization efficiency (KUtE), i.e. the reciprocal of shoot K concentration ($1/[K]_{shoot}$), and K acquisition efficiency (KUpE), i.e. the shoot K content. Both traits showed significant genetic variability within the diversity foundation set of 376 accessions and 74 commercial genotypes of *B. oleracea*. The authors also identified several QTLs associated with shoot K concentration in the AGDH mapping population (Sebastian et al. 2000) grown in the field environment and greenhouse. These loci were significantly correlated with $[K]_{shoot}$ among genotypes grown in both conditions, but differed between environments. One QTL located on chromosome 7 was confirmed using substitution lines. Further studies showed that this region corresponds to chromosome 4 of *A. thaliana*, which includes the *AtKUP9*, *AtAKT2*, *AtKAT2*, and *AtTPK3* K^+ transporter genes. In summary, these studies suggest that QTLs associated with mineral concentration identify genomic regions that are involved with other agronomic traits, such as photosynthetic efficiency and fatty acid profiles (Hall et al. 2005, Barker et al. 2007). On the other hand, planning of breeding programs based on MAS should take into account the experimental conditions, including where crops are grown.

10.3 Transgenesis

Plant transformation is an invaluable technique for understanding gene function via transfer of DNA from any organism into the genome of other species or involves the reduction of expression of specific endogenous genes through gene silencing. Transgenic plants produced in this way are an important resource for both basic research and plant breeding. In *B. oleracea*, this process is accomplished by *Agrobacterium tumefaciens*-mediated transformation methods (De Block et al. 1989; Pius and Achar 2000; Kuginuki and Tsukazaki 2001; Sparrow et al. 2006; Bhalla and Singh 2008). Many studies have shown that key factors, such as susceptibility to *Agrobacterium* and the ability of tissue culture to regenerate, which influence the effectiveness of transformation, are genotype-dependent. To select *B. oleracea* genotypes essential for transformation, Sparrow and coworkers (2015) screened a set of DH lines from the AGDH mapping population (Sebastian et al.

2000). Candidate genotypes were identified using phenotypic markers, i.e. susceptibility to *A. tumefaciens*, shoot regeneration potential, and mode of shoot regeneration. The DH1012 line was selected as the model genotype for transformation and it has been used in further studies (Sparrow et al. 2004b; Sparrow and Irwin 2015). In summary, such features as the high frequency of transgenic plant formation, the speed and efficiency of tissue regeneration, and self-compatibility indicate the potential of the DH1012 line for testing gene function in *Brassica* using the transformation approach.

10.4 Breeding for Glucosinolate Profiles

Glucosinolates (GSLs) are a naturally occurring group of secondary metabolites that are hydrolyzed by the endogenous enzyme myrosinase into isothiocyanates, thiocyanates, and nitriles (Fenwick et al. 1983). Over 200 GSLs have been detected in *Brassica* crops, with several GSLs and about 300 of their compounds in *B. oleracea*. Many studies indicate that the type, content, and profile of GSLs vary in brassicas depending on the cultivar genotype, growth stage, plant organs tested, and environmental conditions, as well as agricultural treatments and post-harvest conditions (Kushad et al. 1999; Kang et al. 2006; Nilsson et al. 2006; Bellostas et al. 2007; Velasco et al. 2007; Pek et al. 2012; Wang et al. 2012; Park et al. 2013; Bhandari and Kwak 2015). Broccoli tissues are particularly rich in GSLs, but their concentrations differ between genotypes and with the seasons (Shapiro et al. 2001; Bhandari and Kwak 2014). The total GSL concentration is highest in seeds and then in sprouts, roots, and shoots (Bhandari and Kwak 2015). The average content of flavonoids in broccoli is higher in the spring than in autumn (Shiva and Jung-Ho 2014). Other studies showed that the production of some GSLs may be dependent on subspecies. For example, glucoerucin (4-methylthiobutyl glucosinolate) is produced in cabbage, but not the edible organs of kale, kohlrabi, and cauliflower (Yi et al. 2015).

GSLs and their products can have different biological functions, including anti-carcinogenic, anti-bacterial, anti-fungal, anti-oxidative, and allelopathic properties (Higdon et al. 2007; Latte et al. 2011). Initially, many studies focused on the identification of GSL decay products with anti-cancer activity (Bhandari and Kwak 2015). Indole-3-carbinol is the best-known anti-cancer compound in *B. oleracea*. It causes the gradual death of tumor cells by acting as an antagonist to estrogen. Other compounds such as isothiocyanates (ITCs) have been shown to have anti-proliferative and antioxidant activity, which ultimately protects cells against the occurrence of oxidative stress (Fahey et al. 2001; Lee and Park 2003; Sonderby et al. 2010). ITCs can prevent cancer in humans and reduce the risk of degenerative diseases (Cartea and Velasco 2008; Van Horn et al. 2008).

Breeding programs are also focused on the identification of genes that control the profile and content of GSLs in *B. oleracea* tissues and the production of brassicas rich in compounds that are beneficial to health. Such programs aim to modify the GSL content so that levels of beneficial GSLs are increased and levels of detrimental GSLs are reduced in the various *Brassica* genotypes. These approaches should lead to the derivation of varieties with higher nutritional value, but also with increased scientific potential, because such varieties will facilitate the study of the biological effects of different levels of specific GSLs in the same genetic background.

Three loci responsible for the profile and content of aliphatic GSLs in *B. oleracea* have been identified (Li et al. 2001). These are *3C-GSL*, which is controlled by a dominant allele of *GSL-PRO*; and *4C-GSL* and *5C-GSL*, which are controlled by a dominant allele of *GSL-ELONG*. Another major gene involved in the synthesis of aliphatic GSLs is *GSL-ALK*, which controls the conversion of methylsulphinyl GSL into alkenyl GSL. In the indolic GSL pathway, the key loci are *CYP79B2*, *CYP79B3*, or *CYP83B*1 (Mikkelsen et al. 2000; Bak et al. 2001; Naur et al. 2003). Recently, six kale genotypes (*B. oleracea* var. *acephala*) have been derived with a high content of sinigrin (HSIN), glucoiberin (HGIB) or glucobrassicin (HGBS) and a low content of sinigrin (LSIN), glucoiberin (LGIB) or glucobrassicin (LGBS) (Sotelo et al. 2016). The results indicate that modifications of SIN and GIB concentrations through selection are associated with the *GSL-ALK* locus. This locus may participate in an indirect response to changes in the GBS concentration. Changes in CYP81F2 expression may also affect GBS content (Sotelo et al. 2016). The authors found that the desired modifications in the content of these GSLs in leaves were reflected in the content of the same GSLs in flower buds and seeds. Clearly, the genes identified are crucial for the modification of the content of the above three GSLs throughout the kale plant.

10.5 Genetics and Breeding of Disease Resistance

Diseases can significantly affect *B. oleracea* production, leading to loss of yield and particularly in detrimental effects on product quality. The main strategies for combating disease rely on practices such as chemical pesticide treatments that can be harmful to the environment and are also not well accepted by vegetable consumers. The search for sources of genetic resistance to disease has become an alternative to chemical or other costly management methods. Disease resistance is a complex process controlled by monogenic and/or polygenic traits with additive effects depending on the plant species. Many breeding programs have searched for sources of resistance within *B. oleracea* germplasms, and include the identification of resistance (R) genes and QTLs associated with specific diseases. The most important diseases developing in vegetable crops are cabbage Fusarium wilt, caused by *Fusarium oxysporum*; clubroot, caused by *Plasmodiophora brassicae*; *Sclerotinia* stem rot, caused by the white mold fungus *Sclerotinia sclerotiorum* (Lib.) de Bary; black rot, caused by the gram-negative bacterium *Xanthomonas campestris* pv. *campestris* (Pammel) Dowson; blackleg, caused by *Leptospheria* species; ringspot, caused by the homothallic ascomycete *Mycosphaerella brassicicola* (Duby) Lindau; white rust, caused by *Albugo candida* (Pers. ex Fr.) O. Kuntze; oomycete downy mildew caused by the oomycete *Hyaloperonospora parasitica* (syn. *Peronospora parasitica*); black spot caused by *Alternaria brassicicola*; and head splitting.

Cabbage Fusarium Wilt (CFW) is one of the most devastating diseases of cabbage, resulting in severe losses in yield and quality (Koike et al. 2003). This pathogen initially infects the cabbage roots, and then colonizes the xylem vessels, leading to leaf wilt and even to wilting of the entire plant. Breeding works have focused on the identification of genes controlling CFW resistance and the development of CFW-resistant cultivars (Jiang et al. 2011; Pu et al. 2012; Lv et al. 2013; 2014a; Xing et al. 2016; Liu et al. 2017) (Table 10.1). Generally, genetic mapping of markers linked to CFW-resistance genes, identification of the putative-resistance genes themselves, as well as MAS in cabbage, have been used (Jiang et al. 2011; Pu et al. 2012; Lv et al. 2014a). These methods have led to an understanding of the inheritance pattern of *Fusarium* resistance and the identification of several resistance genes. The stable sequence characterized amplified region (SCAR) marker S46M48199, which is linked in repulsion to the dominant allele of the *Fusarium* wilt resistance gene *FOC1* at a distance of 2.78 cM, has been developed for MAS (Jiang et al. 2011). The CFW gene *Foc-Bo1* has been mapped to linkage group O7 using both a segregation test and QTL analysis (Pu et al. 2012). Shimizu et al. (2014a, b) showed that *Foc-Bo1* is involved in CFW resistance in broccoli. Other workers mapped two InDel markers, named M10 and A1, which flank the *FOC* resistance gene at genetic distances of 1.2 and 0.6 cM, respectively (Lv et al. 2013). Bol037156, which encodes a putative TIR-NBS-LRR (a toll/interleukin-1 receptor-nucleotide binding site-leucine rich repeat) type R protein, represents a candidate-resistance gene (*FOC1*) that maps to chromosome C06 (Lv et al. 2014a). More recently, changes in expression profiles of genes at the four stages of pathogen infection of *B. oleracea* roots have been analyzed using the high-throughput RNA sequencing

(RNA-seq) method (Xing et al. 2016), resulting in a model for a FOC response network in cabbage. This model proposes the participation of many different signaling pathways including calcium signaling, MAPK signaling, SA-UDPG HR, SA-dependent SAR, JA- and ET-mediated pathways, and lignin biosynthesis.

One of the aims of breeding programs is the production of new disease-resistant varieties based on gene transfer by backcrossing. However, the success of traditional methods of selection of phenotypes mainly depends on environmental conditions and interaction between genotypes and environments; they are also time-consuming. To shorten the time needed to obtain hybrids, a combination of methods-microspore culture, whole-genome analysis, and disease resistance MAS-have been used (Liu et al. 2017). Liu and co-authors crossed two lines, 01–20 (highly susceptible to CFW; an elite line with desirable traits and a high combining ability) and 96–100 (highly resistant to CFW), and developed 230 DH lines. Both parent lines have been re-sequenced and 24 InDel markers have been generated. Using the *FOC1* resistance gene and the Fgr13 SSR marker, screening and phenotypic analysis of individual lines BC_1 and BC_2 have been performed. This selection strategy identified an individual line, named YR01–20, which is homozygous for the *FOC1* allele and has a genomic and phenotypic background similar to the 01–20 line. These data show that this strategy may be useful for the generation of new CFW-resistance varieties in *B. oleracea*.

Clubroot (CR) can be responsible for *Brassica* crop yield losses ranging from 20% in China up to 50% in Canada (Strelkov et al. 2006; Chai et al. 2014). The pathogen attacks the underground parts of plants, leading to gall formation on the root and inhibiting the transport of nutrients and water. As a consequence, plants grow abnormally, resulting in yield loss. It is difficult to prevent CR by traditional methods, such as chemical fungicide application and crop rotation, because the pathogen is able to survive for long periods in soil. Therefore, identification of CR-resistance genes and the development of resistant cultivars is the most desirable breeding strategy for the control of this disease. Many studies have focused on the basic science of resistance to *P. brassicae* in *B. napus* and *B. rapa*. Genetic mapping using various DNA molecular markers together with a MAS strategy has identified numerous genes and QTLs relating to CR (Werner et al. 2007; Neik et al. 2017 for review). Among these, *CRa*, *CRb*, *CRk*, *Crr1*, *Crr3*, and *Crr4* have been mapped on five chromosomes and shown to be involved in resistance to Race 2 in *B. rapa* (Chen et al. 2013a). Although a number of CR QTLs have been mapped in cabbage, including *CR2a*, *CR2b*, *pb-3*,*pb-Bol*, and *PbBo* (Anju)1-4 (Landry et al. 1992; Voorrips et al. 1997; Nagaoka et al. 2010), relatively few R genes have been characterized in the *B. oleracea* genome (Crisp et al. 1989) (Table 10.1). Genetic studies suggest that different combinations of both major and minor R genes participate in resistance to *P. brassicae* in *B. oleracea*, and these are controlled by many QTLs (Diederichsen et al. 2009; Nagaoka et al. 2010). Nagaoka et al. (2010) identified five QTLs for Race 4. Recently, several new QTLs for CR resistance were identified using modern technologies such as GBS and the *Brassica* microarray (Lee et al. 2016; Peng et al. 2018). These approaches allow the construction of high-density genetic maps (with an average marker separation of 1.82 cM, which equates to 0.67 Mb) in a short time. Such maps will increase the precision of QTL region identification. Peng et al. (2018) selected 94 F_2 lines with disease incidence from a cross between CR-resistant and -susceptible lines of *B. oleracea*. Six CR-associated traits were analyzed in this population and genotyped with the 60 K *Brassica* SNP array, including probes from both A and C genomes. Based on data from genetic mapping and the SNP microarray, 23 QTLs for these traits were identified, individually explaining from 6.1 to 17.8% of the phenotypic variation (Table 10.1). Other studies used GBS technology to construct a high-resolution genetic map for *B. oleracea* and then to identify CR-resistance genes (Lee et al. 2016). This map, which includes 43,821 SNPs, isderived from a cross between the resistant- and susceptible-to-CR parental *B. oleracea* var. *capitata* lines,

C1220 and C1176. Three major QTLs, comprising one for Race 2 (*CRQTL-YC*) and two for Race 9 (*CRQTL-GN_1* and *CRQTL-GN_2*) of *P. brassicae*, were located on this map. Their locations coincide with previously identified loci for Race 4 in *B. oleracea* (Nagaoka et al. 2010). Comparison of data from genetic mapping in *B. oleracea* and *B. rapa* indicates differences in the location of CR loci and QTL regions controlling resistance in these species. Further differences in mechanisms responsible for CR have also been observed between genotypes of the individual species. In *B. rapa*, CR is a quantitative and qualitative trait, while it is controlled by many minor QTLs in *B. oleracea* (Sakamoto et al. 2008; Nagaoka et al. 2010; Chen et al. 2013b). However, these data are available for some races of *P. brassicae* only. Therefore, further work should be focused on identifying and cloning the major CR genes controlling resistance to various *P. brassicae* races in the individual genotypes within the *Brassica* genus. This will facilitate an understanding of the relationships between the various CR-resistance genes and allow the development of new breeding materials, as well as CR-resistance genotypes, in *Brassica*.

Sclerotinia stem rot (SSR) can cause up to 80% yield losses in *Brassica* species (Mei et al. 2011). The pathogen usually infects plants as mycelia or airborne ascopores leading to necrotrophic changes in affected tissues (Jamaux et al. 1995). Previous culture practices aimed at preventing SSR were not effective due to the wide host range of the pathogen and its capacity to survive for long periods. The lack of commercial *Brassica* varieties with *Sclerotinia* resistance results from a poor understanding of the *S. sclerotiorum* defense response and molecular mechanisms of resistance. Therefore, the current breeding strategy aims to generate resistant varieties to control this disease. Research has mainly focused on the inheritance of resistance at different developmental stages and the identification of R genes. Disease resistance is a quantitative trait and several QTLs relating to the defense response against *S. sclerotiorum* have been mapped in *B. napus* and *B. oleracea* (Mei et al. 2013; Wu et al. 2013; Wei et al. 2014; 2016). Candidate R genes against *S. sclerotiorum* were also mapped in the *B. napus* genome (Wu et al. 2016). Two of these, belonging to the NBS-LRR gene family, have been located on chromosomes A6 and C7 (Li et al. 2015). Others, representing the CC-NBS-LRR gene family, have been mapped to chromosome A9 (Mei et al. 2013) (Table 10.1). Further molecular studies have analyzed different groups of enzymes involved in resistance to *S. sclerotiorum*: polygalacturonase-inhibiting protein, germin-like proteins, mitogen-activated protein kinases (MAPKs), and WRKY transcription factors (Yang et al. 2009; Rietz et al. 2012; Chen et al. 2013; Wang et al. 2014). In *B. napus*, many QTLs for *S. sclerotiorum* resistance have been identified in the C genome, whose donor is *B. oleracea* (Wu et al. 2013; Li et al. 2015). These have been mapped to chromosomes C02, C04, C06, C07, and C09, and indicate that the C genome of *B. oleracea* is a good source of R genes against this pathogen. Furthermore, the wild *B. oleracea* genotypes, 'C01' and *B. incana*, show increased tolerance against *S. sclerotiorum* (Mei et al. 2011; Mei et al. 2013). These were used as parental lines to develop F_2 populations and then map QTLs for *Sclerotinia* resistance. Eighteen QTLs were identified: 12 for leaf resistance and 6 for stem resistance. Recently, whole-genome transcriptome profiling of the response to *S. sclerotorium* (early infection phase) of the resistant and susceptible *B. oleracea* genotypes was performed (Mei et al. 2016), revealing new pathways involved in resistance. These include elements of Ca^{2+} signaling (*CNGC12*, *CaMs*, and *CDPK*s), which regulates reactive oxygen species (ROS) and cell death; these elements were upregulated and repressed after *S. sclerotorium* inoculation in susceptible and resistant *B. oleracea* genotypes, respectively. Mei and colleagues (2013) have hypothesized that Ca^{2+} signaling promotes apoptosis in the susceptible genotype, while contributing to the inhibition of the expansion of necrophic hyphae in the resistant genotype. Similar differences in expression levels have been observed for some

WRKY transcription factors (two copies of *WRKY11* and *WRKY17*) involved in regulating plant defense (Journot-Catalino et al. 2006).

Black rot disease is one of the most serious diseases in brassicas (Vicente and Holub 2013). Its symptoms are V-shaped chlorotic lesions, leading to loss of both yield and quality of the cabbage head. Many molecular studies aim to identify resistance genes and improve understanding of the mechanisms of resistance to black rot in cabbage. The results of these analyses indicate the participation of diverse immune systems in the response to different pathogenic races of *X. campestris* (*Xcc*). Nine races of *Xcc* that can infect *B. oleracea* have been identified (Tonu et al. 2013). Among these, resistance to races 1 and 4 is not known, in contrast to races 2, 3, and 6, which are less aggressive (Vicente et al. 2001). No black rot-resistant varieties of *B. oleracea* are available. Interestingly, some in-depth research points to the A and B genomes of *Brassica* as sources of black rot resistance (Zhou et al. 1997; Westman et al. 1999; Taylor et al. 2002; Tonguc and Griffiths 2004; Wang et al. 2011, 2016). In *B. carinata* (BC genomes), a single dominant gene, *Xca1bc*, conferring resistance to *Xcc* race 1, has been mapped using intron length polymorphic markers to the B7 chromosome (Tonguc et al. 2003; Sharma et al. 2016). Therefore, the priority for breeding programs is introgression of identified sources of resistance from the A and B genomes into the C genome of *B. oleracea*. The transfer of *Xcc* resistance traits have been made, *inter alia*, from *B. napus* (AC genomes) to broccoli via protoplast fusion, from *B. juncea* (AB genomes) to *B. oleracea* via in vitro embryo rescue and from *B. nigra* to *B. oleracea* via somatic hybridization (Zhou et al. 1997; Tonguc and Griffiths 2004; Wang et al. 2011, 2016). Recently, interspecific hybrids between *B. oleracea* and *Brassica* species have been generated by Sharma and colleagues (2017), who developed hybrids between susceptible cauliflower variety 'Pusa Sharad' and Ethiopian mustard 'NPC-9' using an embryo rescue approach and then derived B_1 backcross progeny. Successful introgression of monogenic resistance to *Xcc* race 1 in the BC_1 population was confirmed using the *ATTAP1* (At1g70610) marker. This population can be used with gene pyramiding strategies to develop durable black rot resistance in cabbages and cauliflowers. Many studies indicate an important role for nucleotide-binding site (NBS)-encoding resistance genes in disease resistance. Recently, expression profiles of 31 NBS-encoding genes (TIR, NBS, LRR, and RPW8 protein domains) have been analyzed in cabbage (*B. oleracea* var. *capitata*) (Afrin et al. 2018). RNA-seq analysis showed them to have differential expression profiles in leaf, root, silique, and stem tissues. Further analyses based on qPCR revealed a higher expression level of some of these genes in resistant compared to susceptible lines.

Blackleg disease is caused by two types of fungal isolate, *Leptosphaeria maculans* (Desmaz.) Ces. & De Not. and *L. biglobosa*, which are highly and weakly virulent species, respectively. Blackleg is one of the most common diseases of the *Brassica* genus, including cabbage and oilseed rape (Rico et al. 2001; Dilmaghani et al. 2009; Piliponyte-Dzikiene et al. 2015). The symptoms of the disease are visible as dark small pycnidia on the leaf tissues and root rot, leading tophoma leaf spots or superficial stem lesions and stem cankers (Dilmaghani et al. 2009; Luo et al. 2018). Numerous studies on the biology, pathogenicity, and control of *Leptosphaeria* have been conducted in *B. oleracea*. Piliponytė-Dzikienė and colleagues (2015) analyzed the occurrence and genetic diversity of *L. maculans* and *L. biglobosa* in *B. oleracea* and *B. napus* in Lithuania. They found that isolates of both fungal species were detected on broccoli and oilseed rape, while only *L. maculans* was present on cabbage. Molecular variance studies showed that populations of *Leptosphaeria* species appearing on different host *Brassica* species could be distinguished genetically. Molecular investigations of *Brassica-Leptosphaeria* interactions are advanced in *B. napus*, *B. rapa,* and *B. juncea*. However, no analyses based on genetic mapping and identification of QTLs for blackleg resistance have been conducted in *B. oleracea*. Recently, Robin and co-workers (2017) tried to understand the molecular mechanism of GSL-mediated defense against *L. moculans* in *B. oleracea*. They analyzed genes involved in the GSL biosynthesis pathway that were induced by two *L. maculans* isolates (03–02 and 00–100 s) and correlated gene expression with changes in GSL profiles in four *B. oleracea* lines (two resistant and two susceptible). The researchers demonstrated a correlation between the type of GSL and the degree of resistance, including an increase in the aliphatic GSLs' glucoiberverin (GIV) and glucoerucin (GER) and the indolic GSLs glucobrassicin (GBS) and neoglucobrassicin (NGBS) associated with complete resistance, an increase in either aliphatic (GIV) or indolic (GBS and MGBS) GSL relating to moderate resistance, and an increase in indolic GBS and NGBS associated with both resistance and susceptibility. Additionally, they found a positive correlation between induction of the *MYB34*, *ST5a*, and *CYP81F2* genes and the content of both GBS and MGBS. These results unambiguously confirmed the relationship between GSL biosynthesis genes and changes in GSL content in cabbage after *L. maculans* infection.

Ringspot spreads in countries with temperate and humid climates. The symptoms of the disease are visible on both sides of the leaf blade as well as on the stems in the form of water-saturated foci surrounded by a chlorotic coating (Koike et al. 2007). To prevent an epidemic of disease among cabbages, breeding strategies have been directed toward analyses of the interaction between host and pathogen as well as the generation of resistant varieties (Bennett and Wallsgrove 1994; Lattanzio et al. 2006). Studies have shown that secondary metabolites are involved in the resistance processes (Kliebenstein 2004; Lattanzio et al. 2006). Abuyusuf and colleagues (2018) investigated GSL profiles, and expression profiles of GSL biosynthesis genes in both ringspot-resistant and -susceptible lines of cabbage after infection with *M. brassicicola*. They found a correlation between the content of aliphatic GIV, GBS, and methoxyglucobrassicin (MGBS) and the type of ringspot-resistant line of cabbage. The content of these GSLs increased in the resistant line and decreased in the susceptible line. Two copies of the *BolMYB34*gene, as well as *BolST5b* and *BolST5c*, were upregulated in the resistant line in line with an accumulation of GIV, GBS, and MGBS. This demonstrates a GSL-mediated defense mechanism against *M. brassicicola* in cabbage.

White rust is caused by *A. candida*, of which 36 races/pathotypes have been identified; race 9 is known to infect *B. oleracea* (Williams 1985; Saharan 2010). The main symptoms shown by infected plants are white pustules on the cotyledons, leaves, and stems, which are surrounded by chlorotic zones (local infection), and also 'staghead' galls in the apical buds and inflorescences (systemic infection) (Verma and Petrie 1980; Kole et al. 2002). The genetic control of resistance to *A. candida* has been studied in *B. napus* and *B. carinata*, and some R genes have been characterized (Scarth et al. 1988; Verma and Bhowmik 1989; Liu and Rimmer 1992; Ferreira et al. 1995b). The sources of resistance to white rust in *B. oleracea* have also been identified by evaluating the reaction to Portuguese isolate Ac502 of *A. candida* within 400 accessions representing the genetic and geographic diversity of this species (Santos and Dias 2004). This study revealed great variability, from complete resistance to full susceptibility, both between and within the set of *B. oleracea* accessions. Sources of resistance have been found among nine accessions including the kales INRA18 and INRA62, the cauliflowers HRI4856, HRI4866, and HRI5424, the loose-head cabbage HRI11555, the savoy cabbage BRA848, the black broccoli HRI6318, and the Portuguese tronchuda cabbage ISA207. These accessions represent useful material for breeding programs targeted at white rust resistance. In further studies, fourteen inbred lines generated from resistant accessions were tested against two Portuguese isolates Ac503 and Ac504 of *A. candida*, as well as a UK isolate. However, no difference was found in the reaction of the *B. oleracea* accessions to the various *A. candida* isolates tested.

Downy mildew is a common disease among *B. oleracea* crops including cabbage, broccoli, cauliflower, kale, and Brussels sprouts. The causative agent, *H. parasitica*, is usually found in close association with the white rust pathogen, *A. candida*, under natural conditions (Bains and Joothy 1985). Downy mildew reveals itself at all developmental stages, and especially in young seedlings and specific organs such as cauliflower curds (Channon 1981). The application of fungicides has only a limited effect against *H. parasitica,* and therefore a significant effort has been made to identify major disease-resistance genes and to develop downy mildew-resistant cultivars of the various *Brassica* crops. Several sources of resistance against *H. parasitica* have been detected in *B. oleracea*. They include single dominant genes in broccoli and cauliflower, a single recessive gene in cauliflower, recessive genes in kale, Savoy cabbage, and Brussels sprouts, as well as multiple (additive) genes in cauliflower, broccoli, and cabbage (Hoser-Krauze et al. 1984, 1995; Moss et al. 1988; Jensen et al. 1999; Farnham et al. 2002; Carlsson et al. 2004). Breeding programs are also striving to develop tools that allow to control the vegetable production for bacterial or fungal infections. Complex approaches have been used to ensure the control of downy mildew for UK vegetable production (Vicente et al. 2012). The authors selected fifteen downy mildew-resistant lines of *B. oleracea*, which were inoculated with 42 *H. parasitica* isolates from the UK collection. Vicente and colleagues found three lines with high isolate-specific resistance (EBH502, EBH508 and EBH527) and three others with a broad-spectrum resistance (EBH525, EBH544 and EBH550). This second group of lines has been crossed with susceptible broccoli Cal18b and cauliflower cv. Lateman lines, aiming for introgression of the broad-spectrum R-resistance gene into non-resistant genotypes. Analysis of the inheritance of this resistance within the resultant populations showed that *H. parasitica* resistance is controlled by a single dominant gene. F_1 generation DH lines were then backcrossed twice to the susceptible parent to produce broccoli and cauliflower plants with resistance in the parental line. These promising studies show that a combination of resistant lines with broad and narrow pathogen-specificity could be used to control downy mildew in UK *Brassica* production.

Alternaria black spot is caused by a wide spectrum of necrotrophic fungal species of the *Alternaria* genus, such as *A. brassicae* (Berk.) Sacc., *A. brassicicola* (Schw.) Wiltsh., *A. raphanin* Groves & Skolko, and *A. alternata* (Fr.) Kreissler. Of these, *A. brassicicola* mainly infects vegetable crops such as *B. oleracea* in warmer climates (Humpherson-Jones and Maude 1982). The main symptoms of the disease include damping-off of seedlings, spotting of leaves, which is visible as black stains with chlorotic yellow halos, blackleg of cabbage heads, and spotting of florets of cauliflowers and broccoli. Disease prevention in plant crops is based on agricultural management practices and fungicidal treatment. There are limited options for strategies that are less aggressive toward the environment due to the lack of *Brassica* accessions resistant to *Alternaria* and a paucity of information about the molecular mechanisms of plant-pathogen interaction. Current research approaches rely on the identification of candidate fungal pathogenicity genes and characterization of the host response by analysis of host-pathogen interactions. A few years ago, the analysis of an *A. brassicicola-B. oleracea* interaction was conducted by screening the EST collection of a suppression subtractive hybridization (SSH) cDNA library derived from pathogen-infected plant leaves (Cramer et al. 2006). This library contains a mixture of both host and fungal genes. The results allowed the identification of candidate *A. brassicicola* genes and of processes involved in the pathogenesis process of *B. oleracea*. Among these, several genes involved in plant invasion and others with lipase activity in the lipid signaling pathway were found. Genes encoding enzymes such as fungal polygalacturonases involved in host penetration/cell-wall degradation were also identified. Another group of genes was those induced by oxidative stress such as catalases, peroxidases, and superoxide dismutases. Analysis of gene profiles in cDNA libraries showed upregulation of transcription factor WRKY33, which is involved in defense processes mediated by salicylic acid. These results shed new light on the complexity of the interaction between pathogen and plant host.

Head-splitting is observed in many fruit and vegetables, such as apple, lychee, chinese cabbage, and tomato (Proctor and Lougheed 1980; Peet 1992; Wang et al. 2006; Zhuang et al. 2009; Niu et al. 2010; Su et al.2012). One of the factors that most likely cause head-splitting is irregular overhead watering; others are light conditions, temperature, and suboptimal humidity (Wang et al. 2006; Niu et al. 2010). Genetic studies on head-splitting resistance have focused on genetic mapping and identification of QTLs. Pang and co-workers (2015) constructed a linkage map in *B. oleracea* using an F_2 population derived from a cross between '748' (head-splitting-resistant inbred line) and '747' (head-splitting-susceptible inbred line). They detected six QTLs relating to resistance to head-splitting on chromosomes C02, C04, and C06 (Table 10.1). The relevant region on chromosome C02 corresponds to chromosome 1 of *A. thaliana*. Analysis of anatomical features of the two parental lines identified three factors associated with head-splitting in cabbage: irregular cellulose microfibrils, markedly larger cell size, and thinner walled epidermis cells. In another study, mixed techniques such as gene/polygene inheritance analysis and QTL mapping have been used to investigate traits for head-splitting resistance (HSR) (Su et al. 2015). In this research, a set of DH lines derived from a cross between head-splitting-susceptible inbred cabbage line 79–156 and resistant line 96–100 were used to construct a genetic map with SSR and InDel markers. Su and co-workers (2015) detected 9 QTLs for HSR located on chromosomes 3, 4, 7, and 9 (Table 10.1). Together, these data should aid the development of breeding strategies based on MAS that will improve our understanding of the genetic regulation of HSR.

10.6 Breeding for Tolerance to Insect Pests

One of the most important insect threats to *Brassica* species is the large white butterfly (LWB) *Pieris brassicae* L., a cosmopolitan insect from the Pieridae family. LWB larvae significantly reduce the fitness of infested plants. One form of plant defense against such pests is secondary metabolic compounds, which are produced and stored in plant tissues and when consumed by herbivores reduce their growth rate, adult size, and survival. Among these are a number of well-characterized GSLs (Rask et al. 2000). In *B. oleracea,* several studies of secondary metabolites, mainly GSLs and flavonoids, have been conducted (Renwick et al. 1992; van Loon et al. 2002; Ferreres et al. 2007). Using HPLC-DAD-MS/MS-ESI, Ferreres and co-authors (2007) identified 20 flavonoids, including 16 kaempferol and four quercetin derivatives, in LWB larvae reared on tronchuda cabbage leaves. Only some of the flavonoids (kaempferol 3-*O*-sophoroside, kaempferol 3-*O*-sophoroside-7-*O*-glucoside and kaempferol 3-*O*-sophoroside-7-*O*-sophoroside) present in the cabbage were also found in the larvae. This indicates that *P. brassicae* has the ability to metabolize other flavonoids contained in cabbage leaves. In later studies, Ferreres and colleagues (2009) explored the metabolism of phenolics in fasted *P. brassicae* larvae and in their excrement, as well as in the host plant *B. oleracea* L. var. *costata* DC. They found that *P. brassicae* sequesters and metabolizes the kale's phenolic compounds. To understand the mechanism of *B. oleracea*-insect interaction, comparative genome-wide gene expression profiling of two white cabbage (*B. oleracea* var. *capitata*) varieties, Rivera and Christmas Drumhead, infested with *P. rapae* caterpillars, was performed using an *A. thaliana* 70-mer oligonucleotide microarray

(Broekgaarden et al. 2007). Both genotypes differed in ability to defend against insect feeding. The expression patterns of the two *B. oleracea* varieties differed in timing (i.e., after 24, 48, and 72 h of caterpillar feeding) as well as quantitatively. Among genes induced only in Rivera were those encoding a putative glutathione S-transferase (GST, At1g27130) and a putative trypsin inhibitor (At2g43520). GSTs belong to a group of stress-response proteins that defend cells against oxidative damage, while trypsin inhibitors protect against the proteolytic enzymes of herbivores. Only one-third of all induced genes were expressed in both varieties. They included genes of the lectin family, which are involved in defense processes by binding glycoconjugates in the intestinal tract of insects, and genes of the tryptophan biosynthesis pathway. This research was extended by an analysis of the jasmonate-induced transcriptional response to *P. rapae* infestation. This showed that *P. rapae* induces more jasmonate-independent than jasmonate-dependent genes. In other work, Das and co-workers (2018) studied the effect of various temperatures on the biology of *P. brassicae* feeding on broccoli leaves under laboratory conditions. They found that 30°C is the most favorable temperature for the development of LWB on broccoli leaves. Thus, various defense mechanisms against *P. rapae* in both *B. oleracea* cultivars have been identified.

10.7 Summary

In *B. oleracea*, many studies and breeding programs aimed at understanding the genetic control of disease are well established. Agronomic approaches focus on the development of methods to allow correct evaluation of disease symptoms and the identification of pathogenicity determinants and fungicides for disease prevention. In addition, the generation of new disease-resistant cultivars is highly desirable on both economic and environmental grounds. Thus, one of the most important strategies in modern plant breeding is the search for new disease-resistance genes; in the case of *B. oleracea*, this includes the possibility of incorporating such genes from the closely related species, *B. napus* and *B. carinata*. So far, these efforts have been time-consuming. Another problem is the rapid evolution of new, virulent pathotypes that then attack resistant cultivars. With the development of whole-genome sequencing and functional genomics-based tools, which is stimulating an improved understanding of the genetic basis of resistance and host-pathogen interaction, progress in the breeding of resistant *B. oleracea* varieties will be facilitated. Especially, a diversity foundation set (DFS) of 376 *B. oleracea* accessions including many common allelic variation, mutant collections, informative genome sequencing databases, and techniques for genetic transformation will become helpful. Indeed, in recent years, many papers have described the use of these tools, but the main problems remain the homozygosity of cultivar lines and the limited transfer of knowledge and techniques between the laboratories of researchers and plant breeders. For example, numerous genetic maps have been developed using different morphotypes with various marker types; different nomenclatures are used for the same chromosomes and genes; and different genetic materials and tools are used in different laboratories. Clearly, this does not facilitate the transfer of knowledge between groups. Therefore, a major challenge for the field is to establish a platform for the integration of data. For Brassica research, most genetic data and tools are collected at the www.brassica.info website. They include data in the field of genetics and genomics, such as registries of reference datasets and nomenclature standards (genes, markers, loci), as well as links to different *Brassica* accessions. While these are excellent first steps, the Brassica research community needs to continue to work toward the full integration of resources and nomenclature. Such integration will likely accelerate progress in research and breeding, as it has in organisms such as *A. thaliana*.

References

Abuyusuf Md, Robin AH, Kim HT, Islam MD, Park J, Nou IS (2018) Altered glucosinolate profiles and expression of glucosinolate biosynthesis genes in ringspot-resistant and susceptible cabbage lines. Intl J Mol Sci 19:2833

Afrin KS, Rahim MA, Park JI, Natarajan S, Kim HT, Nou IS (2018) Identification of NBS-encoding genes linked to black rot resistance in cabbage (*Brassica oleracea* var. *capitata*). Mol Biol Rep 45:773

Bains SS, Joothy JS (1985) Association of *Peronospora parasitica* with *Albugo candida* on *Brassica juncea* leaves. J Phytopathol 112:28–31

Bak S, Feyereisen R, Tax FE, Feldmann KA, Galbraith DW (2001) *CYP83B1*, a cytochrome P450 at the metabolic branch point in auxin and indole glucosinolate biosynthesis in *Arabidopsis*. Plant Cell 13:101–111

Barker GC, Larson TR, Graham IA, Lynn JR, King GJ (2007) Novel insights into seed fatty acid synthesis and modification pathways from genetic diversity and quantitative trait loci analysis of the *Brassica* C genome. Plant Physiol 144:1827–1842

Bastow R, Mylne JS, Lister C, Lippman Z, Martienssen RA, Dean C (2004) Vernalization requires epigenetic silencing of *FLC* by histone methylation. Nature 427:164–167

Bellostas N, Kachlicki P, Sorensen JC, Sorensen H (2007) Glucosinolate profiling of seeds and sprouts of *B. oleracea* varieties used for food. Sci Hort 114:234–242

Bennett RN, Wallsgrove RM (1994) Secondary metabolites in plant defence mechanisms. New Phytol 127:617–633

Bhalla PL, Singh MB (2008) *Agrobacterium*-mediated transformation of *Brassica napus* and *Brassica oleracea*. Nat Protoc 3:181–189

Bhandari SR, Kwak JH (2014) Seasonal variation in phytochemicals and antioxidant activities in different tissues of various broccoli cultivars. Afr J Biotechnol 13:604–615

Bhandari SR, Kwak JH (2015) Chemical composition and antioxidant activity in different tissues of *Brassica* vegetables. Molecules 20:1228–1243

Blümel M, Dally N, Jung C (2015) Flowering time regulation in crops-what did we learn from *Arabidopsis*? Curr Opin Biotechnol 32:121–129

Bohuon EJR, Ramsay LD, Craft JA, Arthur AE, Marshall DF, Lydiate DJ, Kearsey MJ (1998) The association of flowering time quantitative trait loci with duplicated regions and candidate loci in *Brassica oleracea*. Genetics 150:393–401

Bothmer R, Gustafsson M, Snogerup S (1995) *Brassica* sect. *Brassic*a (Brassicaceae). II. Inter- and intraspecific crosses with cultivars of *B. oleracea*. Genet Resour Crop Evol 42:165–178

Broadley MR, Bowen HC, Cotterill HL, Hammond JP, Meacham MC, Mead A, White PJ (2004) Phylogenetic variation in the shoot mineral concentration of angiosperms. J Exp Bot 55:321–336

Broadley MR, Hammond JP, King GJ, Astley D, Bowen HC, Meacham MC, Mead A, Pink DA, Teakle GR, Hayden RM, Spracklen WP, White PJ (2008) Shoot calcium and magnesium concentrations differ between subtaxa, are highly heritable, and associate with potentially pleiotropic loci in *Brassica oleracea*. Plant Physiol 146:1707–1720

Broadley MR, Lochlainn S, Hammond J, Bowen H, Cakmak I, Eker S, Erdem H, King G, White P (2010) Shoot zinc (Zn) concentration varies widely within *Brassica oleracea* L. and is affected by soil Zn and phosphorus (P) levels. J Hort Sci Biotechnol 5:375–380

Broekgaarden C, Poelman EH, Steenhuis G, Voorrips RE, Dicke M, Vosman B (2007) Genotypic variation in genome-wide transcription profiles induced by insect feeding: *Brassica oleracea*-*Pieris rapae* interactions. BMC Genom 8:239

Camargo LEA, Osborn TC (1996) Mapping loci controlling flowering time in *Brassica oleracea*. Theor Appl Genet 92:610–616

Carlsson M, Von Bothmer R, Merker A (2004) Screening and evaluation of resistance to downy mildew (*Peronospora parasitica*) and clubroot (*Plasmodiophora brassicae*) in genetic resources of *Brassica oleracea*. Hereditas 141:293–300

Cartea ME, Velasco P (2008) Glucosinolates in *Brassica* foods: bioavailability in food and significance for human health. Phytochem Rev 7:213–229

Chai AL, Xie XW, Shi YX, Li BJ (2014) Research status of clubroot (*Plasmodiophora brassicae*) on cruciferous crops in China. Can J Plant Pathol 36:142–153

Channon AG (1981) Downy mildew of brassicas. In: Spencer DM (ed) The Downy Mildews. Academic Press, London, UK, pp 321–339

Chen C, Zhuang M, Fang Z-Y, Wang Q, Zhang Y, Liu Y-M, Yang L-M, Cheng F (2013a) A co-dominant marker BoE332 applied to marker-assisted selection of homozygous male-sterile plants in cabbage (*Brassica oleracea* var. *capitata* L.). J Integr Agri 12:596–602

Chen J, Jing J, Zhan Z, Zhang T, Zhang C, Piao Z (2013a) Identification of novel QTLs for isolate-specific partial resistance to *Plasmodiophora brassicae* in *Brassica rapa*. PLOS One 8:e85307

Chen X, Liu J, Lin G, Wang A, Wang Z, Lu G (2013b) Overexpression of *AtWRKY28* and *AtWRKY75* in *Arabidopsis* enhances resistance to oxalic acid and *Sclerotinia sclerotiorum*. Plant Cell Rep 32:1589–1599

Cramer RA, LA Rota CM, Cho Y, Thon M, Craven KD, Knudson DL, Mitchell TK, Lawrence CB (2006) Bioinformatic analysis of expressed sequence tags derived from a compatible *Alternaria brassicicola*-*Brassica oleracea* interaction. Mol Plant Pathol 7:113–124

Crisp P, Crute I, Sutherland R, Angell SM, Bloor K, Burgess H, Gordon PL (1989) The exploitation of genetic resources of *Brassica oleracea* in breeding for resistance to clubroot (*Plasmodiophora brassicae*). Euphytica 42:215–226

Das P, Mishra RC, Chauhan D, Kumar V, Bahuguna P (2018) Response of different temperatures on biology of cabbage butterfly, *Pieris brassicae* (L.), Lepidoptera: Pieridae on broccoli leaves in laboratory. Intl J Pure App Biosci 6:1244–1251

De Block M, De Brouwer D, Tenning P (1989) Transformation of *Brassica napus* and *Brassica oleracea* using *Agrobacterium tumefaciens* and the expression of the bar and neo genes in the transgenic plants. Plant Physiol 91:694–701

Diederichsen E, Frauen M, Linders EG, Hatakeyama K, Hirai M (2009) Status and perspectives of clubroot resistance breeding in crucifer crops. J Plant Growth Regul 28:265–281

Dilmaghani A, Balesdent MH, Didier JP, Wu C, Davey J, Barbetti MJ, Li H, Moreno-Rico O, Phillips D, Despeghel JP, Vincenot L, Gout L, Rouxel T (2009) The *Leptosphaeria maculans* - *Leptosphaeria biglobosa* species complex in the American continent. Plant Pathol 58:1044–1058

Divaret I, Margalé E, Thomas G (1999) RAPD markers on seed bulks efficiently assess the genetic diversity of a *Brassica oleracea* L. collection. Theor Appl Genet 98:1029–1035

Fahey JW, Zalcmann AT, Talalay P (2001) The chemical diversity and distribution of glucosinolates and isothiocyanates among plants. Phytochemistry 56:5–51

Farnham MW, Wang M, Thomas CE (2002) A single dominant gene for downy mildew resistance in broccoli. Euphytica 128:405–407

Fenwick GR, Heaney RK, Mullin WJ (1983) Glucosinolates and their breakdown products in food and food plants. Crit Rev Food Sci Nutr 18:123–201

Ferreira ME, Satagopan J, Yandell BS, Williams PH, Osborn TC (1995a) Mapping loci controlling vernalisation requirement and flowering time in *Brassica napus*. Theor Appl Genet 90:727–732

Ferreira ME, Williams PH, Osborn TC (1995b) Mapping of locus controlling resistance to *Albugo candida* in *Brassica napus* using molecular markers. Phytopathology 85:218–220

Ferreres F, Sousa C, Valentão P, Pereira JA, Seabra RM, Andrade PB (2007) Tronchuda cabbage flavonoids uptake by *Pieris brassicae*. Phytochemistry 68:361–367

Ferreres F, Fernandes F, Oliveira JM, Valentão P, Pereira JA, Andrade PB (2009) Metabolic profiling and biological capacity of *Pieris brassicae* fed with kale (*Brassica oleracea* L. var. *acephala*). Food Chem Toxicol 47:1209–1220

Golicz A, Bayer PE, Barker GC, Edger P, Kim H, Martinez PA, Kit C, Chan K, Severn-Ellis A, McCombie WR, Parkin IAP, Paterson AH, Pires JC, Sharpe AG, Tang H, Teakle GR, Town CD, Batley J, Edwards D (2016) The pangenome of an agronomically important crop plant *Brassica oleracea*. Nat Commun 7:13390

Halkier BA, Gershenzon J (2006) Biology and biochemistry of glucosinolates. Annu Rev Plant Biol 57:303–333

Hall NM, Griffiths H, Corlett JA, Jones HG, Lynn J, King GJ (2005) Relationships between water-use traits and photosynthesis in *Brassica oleracea* resolved by quantitative genetic analysis. Plant Breed 124:557–564

Higdon JV, Delage B, Williams DE, Dashwood RH (2007) Cruciferous vegetables and human cancer risk: epidemiologic evidence and mechanistic basis. Pharmacol Res 55:224–236

Holme IB, Torp AM, Hansen LN, Andersen SB (2004) Quantitative trait loci affecting plant regeneration from protoplasts of *Brassica oleracea*. Theor Appl Genet 108:1513–1520

Hoser-Krauze J, Łąkowska-Ryk E, Antosik J (1984) Resistance of cauliflower and broccoli (*B. oleracea* L. *botrytis* L.) seedlings to downy mildew. *Peronospora parasitica*. Cruciferae Newsletter 9:3–92

Hoser-Krauze J, Lakowska-Ryk E, Antosik J (1995) The inheritance of resistance of some *Brassica oleracea* L. cultivars and lines to downy mildew, *Peronospora parasitica* (Pers) ex. Fr. J Appl Genet 36:27–33

Humpherson-Jones FM, Maude RB (1982) Studies on the epidemiology of *Alternaria brassicicola* in *Brassica oleracea* seed production crops. Ann Appl Biol 100:61–71

Irwin JA, Soumpourou E, Lister C, Ligthart J-D, Kennedy S, Dean C (2016) Nucleotide polymorphism affecting *FLC* expression underpins heading date variation in horticultural brassicas. Plant J 87:597–605

Jamaux J, Gelie B, Lamarque C (1995) Early stages of infection of rapeseed petals and leaves by *Sclerotinia sclerotiorum* revealed by scanning electron microscopy. Plant Pathol 44:22–30

Jensen BD, Hockenhull J, Munk L (1999) Seedling and adult plant resistance to downy mildew (*Peronospora parasitica*) in cauliflower (*Brassica oleracea* convar. *botrytis* var. *botrytis*). Plant Pathol 48:604–612

Jiang M, Zhao Y, Xie J, Tian R, Chen Y, Kang J (2011) Development of a SCAR marker for Fusarium wilt resistance in cabbage. Sci Agri Sin 44:3053–3059 (in Chinese)

Journot-Catalino N, Somssich IE, Roby D, Kroj T (2006) The transcription factors WRKY11 and WRKY17 act as negative regulators of basal resistance in *Arabidopsis thaliana*. Plant Cell 18:3289–3330

Kang JY, Ibrahim KE, Juvik JA, Kim DH, Kang WH (2006) Genetic and environmental variation of glucosinolate content in Chinese cabbage. Hort Sci 41:1382–1385

Kennard WC, Slocum MK, Figdore SS, Osborn TC (1994) Genetic analysis of morphological variation in *Brassica oleracea* using molecular markers. Theor Appl Genet 87:721–732

Kliebenstein DJ (2004) Secondary metabolites and plant/environment interactions: A view through

Arabidopsis thaliana tinged glasses. Plant Cell Environ 27: 675–684

Koike ST, Subbarao KV, Davis RM, Turini TA (2003) Vegetable diseases caused by soilborne pathogens. ANR Publication, University of California, California, CA

Koike ST, Gladders P, Paulus A (2007) Vegetable Diseases. Academic Press, New York, A colour handbook

Kole C, Williams PH, Rimmer SR, Osborn TC (2002) Linkage mapping of genes controlling resistance to white rust (*Albugo candida*) in *Brassica rapa* (syn. *campestris*) and comparative mapping to *Brassica napus* and *Arabidopsis thaliana*. Genome 45:22–27

Kopsell DA, Kopsell DE (2006) Accumulation and bioavailability of dietary carotenoids in vegetable crops. Trends Plant Sci 11:499–507

Kuginuki Y, Tsukazaki H (2001) Regeneration ability and *Agrobacterium*-mediate transformation of different cultivars in *Brassica oleracea* L. and *B. rapa*. J Jap Soc Hort Sci 70:682–690

Kushad MM, Brown AF, Kurilich AC, Juvik JA, Klein BP, Wallig MA, Jeffery EH (1999) Variation of glucosinolates in vegetable crops of *Brassica oleracea*. J Agri Food Chem 47:1541–1548

Lan TH, Paterson AH (2001) Comparative mapping of QTLs determining the plant size of *Brassica oleracea*. Theor Appl Genet 103:383–397

Landry BS, Hubert N, Crete R, Chang MS, Lincoln SE, Etoh T (1992) A genetic map for *Brassica oleracea* based on RFLP markers detected with expressed DNA sequences and mapping of resistance genes to race 2 of *Plasmodiophora brassicae* (Woronin). Genome 35:409–420

Lattanzio V, Lattanzio VM, Cardinali A (2006) Role of phenolics in the resistance mechanisms of plants against fungal pathogens and insects. Phytochem Adv Res 661:23–67

Latte KP, Appel KE, Lampen A (2011) Health benefits and possible risks of broccoli - an overview. Food Chem Toxicol 49:3287–3309

Lazaro A, Aguinagalde I (1998) Genetic diversity of *Brassica oleracea* L. (Cruciferae) and wild relatives ($2n = 18$) using RAPD markers. Ann Bot 82:829–833

Lee BM, Park KK (2003) Beneficial and adverse effects of chemopreventive agents. Mutation Res 523:265–278

Lee J, Izzah NK, Choi BS, Joh HJ, Lee SC, Perumal S, Seo J, Ahn K, Jo EJ, Choi GJ, Nou IS, Yu Y, Yang TJ (2016) Genotyping-by-sequencing map permits identification of clubroot resistance QTLs and revision of the reference genome assembly in cabbage (*Brassica oleracea* L.). DNA Res 23:29–41

Li G, Riaz A, Goyal S, Abel S, Quiros CF (2001) Inheritance of three major genes involved in the synthesis of aliphatic glucosinolates in *Brassica oleracea*. J Amer Soc Hort Sci 126:427–431

Li J, Zhao Z, Hayward A, Cheng H, Fu D (2015) Integration analysis of quantitative trait loci for resistance to *Sclerotinia sclerotiorum* in *Brassica napus*. Euphytica 205:483–489

Lin SI, Wang JG, Poon SY, Su CL, Wang SS, Chiou TJ (2005) Differential regulation of *FLOWERING LOCUS C* expressed by vernalization in cabbage and *Arabidopsis*. Plant Physiol 137:1037–1048

Liu Q, Rimmer SR (1992) Inheritance of resistance in *Brassica napus* to an Ethiopian isolate of *Albugo candida* from *Brassica carinata*. Can J Plant Pathol 14:116–120

Liu X, Han F, Kong C, Fang Z, Yang L, Zhang Y, Zhuang M, Liu Y, Li Z, Lv H (2017) Rapid introgression of the *Fusarium* wilt resistance gene into an elite cabbage line through the combined application of a microspore culture, genome background analysis, and disease resistance-specific marker assisted foreground selection. Front Plant Sci 8:354

Luo T, Cai X, Li G, Yang L (2018) First report of *Leptosphaeria biglobosa* causing blackleg on broccoli (*Brassica oleracea* var. *italica*) in Central China. Plant Dis 102:1178

Lv H, Yang L, Kang J, Wang Q, Wang X, Fang Z, liu Y, Zhuang M, Zhang Y, Lin Y, Yang Y, Xie B, Liu B, Liu J (2013) Development of InDel markers linked to Fusarium wilt resistance in cabbage. Mol Breed 32:961–967

Lv HH, Fang ZY, Yang LM, Zhang YY, Wang QB, Liu YM, Zhuang M, Yang Y, Xie B, Liu B, Liu J, Kang J, Wang X (2014a) Mapping and analysis of a novel candidate *Fusarium* wilt resistance gene *FOC1* in *Brassica oleracea*. BMC Genom 15:1094

Lv H, Wang Q, Zhang Y, Yang L, Fang Z, Wang X, Liu Y, Zhuang M, Lin Y, Yu H, Liu B (2014b) Linkage map construction using InDel and SSR markers and QTL analysis of heading traits in cabbage. Mol Breed 34:87–98

Lv H, Wang Q, Liu X, Han F, Fang Z, Yang L, Zhuang M, Liu Y, Li Z, Zhang Y (2016) Whole-genome mapping reveals novel QTL clusters associated with main agronomic traits of cabbage (*Brassica oleracea* var. *capitata* L.). Front Plant Sci 7:989

Mei J, Qian L, Disi JO, Yang X, Li Q, Li J, Frauen M, Cai D, Qian W (2011) Identification of resistant sources against *Sclerotinia sclerotiorum* in *Brassica* species with emphasis on *B. oleracea*. Euphytica 177:393–399

Mei J, Ding Y, Lu K, Wei D, Liu Y, Disi JO, Li J, Liu L, Liu S, McKay J, Qian W (2013) Identification of genomic regions involved in resistance against *Sclerotinia sclerotiorum* from wild *Brassica oleracea*. Theor Appl Genet 126:549–556

Mei J, Ding Y, Li Y, Tong C, Du H, Yu Y, Wan H, Xiong Q, Yu J, Liu S, Li J, Qian W (2016) Transcriptomic comparison between *Brassica oleracea* and rice (*Oryza sativa*) reveals diverse modulations on cell death in response to *Sclerotinia sclerotiorum*. Sci Rep 6:33706

Mikkelsen MD, Wittstock U, Hansen CH, Halkier BA (2000) Cytochrome P450 CYP79B2 from *Arabidopsis* catalyzes the conversion of tryptophan to indole-3-acetaldoxime, a precursor of indole glucosinolates and indole-3-acetic acid. J Biol Chem 275:33712–33717

Moreno-Rico O, Frias-Treviño AG, Luna-Ruiz JJ, Manzano-Flores DE, Romero-Cova S, Séguin-Swartz G (2001) Characterization and pathogenicity of isolates of *Leptosphaeria maculans* from Aguascalientes and Zacatecas Mexico1. Can J Plant Pathol 23:270–278

Moss NA, Crute IR, Lucas JA, Gordon PL (1988) Requirements for analysis of host-species specificity in *Peronospora parasitica* (downy mildew). Cruciferae Newsl 13:114–492

Nagaoka T, Doullah MA, Matsumoto S, Kawasaki S, Ishikawa T, Hori H, Okazaki K (2010) Identification of QTLs that control clubroot resistance in *Brassica oleracea* and comparative analysis of clubroot resistance genes between *B. rapa* and *B. oleracea*. Theor Appl Genet 120:1335–1346

Naur P, Petersen B, Mikkelsen M, Bak S, Rasmussen H, Olsen C, Halkier BA (2003) CYP83A1 and CYP83B1, two nonredundant cytochrome P450 enzymes metabolizing oximes in the biosynthesis of glucosinolates in *Arabidopsis*. Plant Physiol 133:63–72

Neik TX, Barbetti MJ, Batley J (2017) Current status and challenges in identifying disease resistance genes in *Brassica napus*. Front Plant Sci 8:1788

Nilsson J, Olsson K, Engqvist G, Ekvall J, Olsson M, Nyman M, Akesson B (2006) Variation in the content of glucosinolates, hydrocinnamic acids, carotenoids, total antioxidant capacity and low-molecular-weight carbohydrates in *Brassica* vegetables. J Sci Food Agri 86:528–538

Niu N, Zheng C, Zhang L, Xu Y, Zhang L, Fu W (2010) Inheritance and RAPD marker of head-splitting trait in Chinese cabbage. China Veg 14:44–48

Okazaki K, Sakamoto K, Kikuchi R, Saito A, Togashi E, Kuginuki Y, Matsumoto S, Hirai M. (2007) Mapping and characterization of *FLC* homologs and QTL analysis of flowering time in *Brassica oleracea*. Theor Appl Genet 114:595–608

Pagliaro B, Santolamazza C, Simonelli F, Rubattu S (2015) Phytochemical compounds and protection from cardiovascular diseases: A state of the art. Biomed Res Intl 2015:918069

Pang W, Li X, Choi SR, Nguyen VD, Dhandapani V, Kim YY, Ramchiary N, Kim JG, Edwards D, Batley J, Na J, Kim HR, Lim YP (2015) Mapping QTLs of resistance to head splitting in cabbage (*Brassica oleracea* L. var. *capitata* L). Mol Breed 35:126

Park MH, Arasu MV, Park NY, Choi YJ, Lee SW, Al-Dhabi NA, Kim JB, Kim SJ (2013) Variation of glucoraphanin and glucobrassicin: Anticancer components in *Brassica* during processing. Food Sci Technol Camp 33:624–631

Peet MM (1992) Fruit cracking in tomato. Hort Technol 2:216–223

Pek Z, Daood H, Nagyne MG, Berki M, Tothne MM, Nemenyi A, Helyes L (2012) Yield and phytochemical compounds of broccoli as affected by temperature, irrigation, and foliar sulfur supplementation. Hort Sci 47:1646–1652

Peng L, Zhou L, Li Q, Wei D, Ren X, Song H, Mei J, Si J, Qian W (2018) Identification of quantitative trait loci for clubroot resistance in *Brassica oleracea* with the use of *Brassica* SNP microarray. Front Plant Sci 9:822

Piliponyte-Dzikiene A, Andriunaite E, Petraitiene E, Brazauskiene I, Statkeviciute G, Brazauskas G (2015) Genetic diversity and occurrence of *Leptosphaeria* spp. on *Brassica oleracea* and *B. napus* in Lithuania. J Plant Pathol 97:265–271

Pius PK, Achar PN (2000) *Agrobacterium*-mediated transformation and plant regeneration of *Brassica oleracea* var. *capitata*. Plant Cell Rep 19:888–892

Proctor JTA, Lougheed EC (1980) Cracking of Golden Russet apples. Can Plant Dis Surv 60:55–58

Pu Z, Motoki S, Zhang Y, Tomohiko N, Takeshi H, Hidetaka H, Satoru M, Ryo F, Keiichi O (2012) Genetic mapping of a *Fusarium* wilt resistance gene in *Brassica oleracea*. Mol Breed 30:809–818

Rae AM, Howell EC, Kearsey MJ (1999) More QTL for flowering time revealed by substitution lines in *Brassica oleracea*. Heredity 83:586–596

Rask L, Andréasson E, Ekbom B, Eriksson S, Pontoppidan B, Meijer J (2000) Myrosinase: gene family evolution and herbivore defense in Brassicaceae. Plant Mol Biol 42:93–113

Razi H, Howell E, Newbury H, Kearsey M (2008) Does sequence polymorphism of *FLC* paralogues underlie flowering time QTL in *Brassica oleracea*? Theor Appl Genet 116:179–192

Renwick JAA, Radke CD, Sachdev-Gupta K, Städler E (1992) Leaf surface chemicals stimulating oviposition by *Pieris rapae* (Lepidoptera: Pieridae) on cabbage. Chemoecol 3:33–38

Ridge S, Brown PH, Hecht V, Driessen RG, Weller JL (2015) The role of *BoFLC2* in cauliflower (*Brassica oleracea* var. *botrytis* L.) reproductive development. J Exp Bot 66:125–135

Rietz S, Bernsdorff FEM, Cai DG (2012) Members of the germin-like protein family in *Brassica napus* are candidates for the initiation of an oxidative burst that impedes pathogenesis of *Sclerotinia sclerotiorum*. J Exp Bot 63:5507–5519

Robin AHK, Yi GE, Laila R, Hossain MR, Park JI, Kim HR, Nou IS (2017) *Leptosphaeria maculans* alters glucosinolate profiles in blackleg disease-resistant and -susceptible cabbage lines. Front Plant Sci 8:1769

Saharan GS (2010) Analysis of genetic diversity in *Albugo*-Crucifer system. Mycol Plant Pathol 40:1–13

Sakamoto K, Saito A, Hayashida N, Taguchi G, Matsumoto E (2008) Mapping of isolate-specific QTLs for clubroot resistance in Chinese cabbage (*Brassica rapa* L. ssp. *pekinensis*). Theor Appl Genet 117:759–767

Santos M, Dias J (2004) Evaluation of a core collection of *Brassica oleracea* accessions for resistance to white rust of crucifers (*Albugo candida*) at the cotyledon stage. Genet Resour Crop Evol 51:713

Scarth R, McVetty PBE, Rimmer SR, Stefansson BR (1988) Stellar low linolenic-high linoleic acid summer rape. Can J Plant Sci 68:509–511

Schiessl SV, Huettel B, Kuehn D, Reinhardt R, Snowdon RJ (2017) Flowering time gene variation in *Brassica* species shows evolutionary principles. Front Plant Sci 8:1742

Schranz ME, Quijada P, Sung S-B, Lukens L, Amasino R, Osborn TC (2002) Characterization and effects of the replicated flowering time gene *FLC* in *Brassica rapa*. Genetics 162:1457–1468

Sebastian RL, Howell EC, King GJ, Marshall DF, Kearsey MJ (2000) An integrated AFLP and RFLP *Brassica oleracea* linkage map from two morphologically distinct doubled-haploid mapping populations. Theor Appl Genet 100:75–81

Sebastian RL, Kearsey MJ, King GJ (2002) Identification of quantitative trait loci controlling developmental characteristics of *Brassica oleracea* L. Theor Appl Genet 104:601–609

Shah MA, Sarker MMR, Gousuddin M (2016) Antidiabetic potential of *Brassica oleracea* var. *italica* in type 2 diabetic Sprague dawley (sd) rats. Intl J Phar Phytochem Res 8:462–469

Shapiro TA, Fahey JW, Wade KL, Stephenson KK, Talalay P (2001) Chemoprotective glucosinolates and isothiocyanates of broccoli sprouts: metabolism and excretion in humans. Cancer Epidemiology, Biomarkers & Prevention: A Publication of the American Association for Cancer Research, Cosponsored by the American Society of Preventive Oncology 10:501–508

Sharma BB, Kalia P, Yadava DK, Singh D, Sharma TR (2016) Genetics and molecular mapping of black rot resistance locus *Xca1bc* on Chromosome B-7 in Ethiopian Mustard (*Brassica carinata* A. *Braun*). PLOS One 11:e0152290

Sharma BB, Kalia P, Singh D, Sharma TR (2017) Introgression of black rot resistance from *Brassica carinata* to cauliflower (*Brassica oleracea botrytis* group) through embryo rescue. Front Plant Sci 8:1255

Shimizu M, Fujimoto R, Ying H, Pu Z, Ebe Y, Kawanabe T, Saeki N, Taylor JM, Kaji M, Dennis ES, Okazaki K (2014a) Identification of candidate genes for Fusarium yellows resistance in Chinese cabbage by differential expression analysis. Plant Mol Biol 85: 247–257

Shimizu M, Pu Z, Kawanabe T, Kitashiba H, Matsumoto S, Ebe Y, Sano M, Funaki T, Fukai E, Fujimoto R, Okazaki K (2014b) Map-based cloning of a candidate gene conferring Fusarium yellows resistance in *Brassica oleracea*. Theor Appl Genet 128:19–130

Smith LB, King GJ (2000) The distribution of Bo-CAL-a alleles in *Brassica oleracea* is consistent with a genetic model for curd development and domestication of the cauliflower. Mol Breed 6:603–613

Sonderby IE, Geu-Flores F, Halkier BA (2010) Biosynthesis of glucosinolates gene discovery and beyond. Trends Plant Sci 15:283–290

Sotelo T, Velasco P, Soengas P, Rodríguez VM, Cartea ME (2016) Modification of leaf glucosinolate contents in *Brassica oleracea* by divergent selection and effect on expression of genes controlling glucosinolate pathway. Front Plant Sci 15:1012

Sparrow PAC, Townsend TM, Arthur AE, Dalle PJ, Irwin JA (2004a) Genetic analysis of *Agrobacterium tumefaciens* susceptibility in *Brassica oleracea*. Theor Appl Genet 108:644–650

Sparrow PAC, Dale PJ, Irwin JA (2004b) The use of phenotypic markers to identify *Brassica oleracea* genotypes for routine high-throughput *Agrobacterium*-mediated transformation. Plant Cell Rep 23:64–70

Sparrow PAC, Dale PJ, Irwin JA (2006) *Brassica oleracea*. Methods Mol Biol 343:417–426

Sparrow PACH, Irwin JA (2015) *Brassica oleracea* and *B. napus*. In: Wang K. (eds) *Agrobacterium* Protocols. Methods in Molecular Biology, vol 1223. Springer, New York, NY

Stansell Z, Hyma K, Fresnedo-Ramírez J, Sun Q, Mitchell S, Björkman T, Hua J (2018) Genotyping-by-sequencing of *Brassica oleracea* vegetables reveals unique phylogenetic patterns, population structure and domestication footprints. Hort Res 5:38

Strelkov SE, Tewari JP, Smith-Degenhardt E (2006) Characterization of *Plasmodiophora brassicae* populations from Alberta, Canada. Can J Plant Pathol 28:467–474

Su YB, Liu YM, Fang ZY, Yang LM, Zhuang M, Zhang YY, Zhang XL, Sun PT (2012) Genetic analysis of head-splitting resistance traits in cabbage. Acta Hort Sin 39:1482–1490

Su Y, Liu Y, Li Z, Fang Z, Yang L, Zhuang M, Zhang Y (2015) QTL analysis of head splitting resistance in cabbage (*Brassica oleracea* L. var. *capitata*) using SSR and InDel makers based on whole-genome re-sequencing. PLOS One 10:e0138073

Taylor JD, Conway S, Roberts SJ, Astley D, Vicente JG (2002) Sources and origin of resistance to *Xanthomonas campestris* pv. *campestris* in *Brassica* genomes. Phytopathology 92:105–111

Thorwarth P, Yousef EAA, Schmid KJ (2018) Genomic prediction and association mapping of curd-related traits in genebank accessions of cauliflower. G3 (Bethesda) 8:707–718

Tonguc M, Griffiths PD (2004) Development of black rot resistant interspecific hybrids between *Brassica oleracea* L. cultivars and *Brassica* accession A 19182, using embryo rescue. Euphytica 136:313–318

Tonguc M, Earle ED, Griffiths PD (2003) Segregation distortion of *Brassica carinata* derived black rot resistance in *Brassica oleracea*. Euphytica 134:269–276

Tonu NN, Doullah MA, Shimizu M, Karim MM, Kawanabe T, Fujimoto R, Okazaki K (2013) Comparison of positions of QTLs conferring resistance to *Xanthomonas campestris* pv. *campestris* in *Brassica oleracea*. Amer J Plant Sci 4:11–20

Uptmoor R, Li J, Schrag T, Stützel H (2012) Prediction of flowering time in *Brassica oleracea* using a quantitative trait loci-based phenology model. Plant Biol 14:179–189

Van Horn L, McCoin M, Kris Etherton P, Burke F (2008) The evidence for dietary prevention and treatment of cardiovascular disease. J Amer Diet Assoc 108:287–331

Van Loon JJ, Wang CZ, Nielsen JK, Gols R, Qiu YT (2002) Flavonoids from cabbage are feeding stimulants for diamondback moth larvae additional to glucosinolates: chemoreception and behaviour. Entomol Exp App. 104:27–34

Velasco P, Cartea ME, Gonzalez C, Vilar M, Ordás A (2007) Factors affecting the glucosinolate content of kale (*Brassica oleracea acephala* group). J Agri Food Chem 55:955–962

Verma PR, Petrie GA (1980) Effect of seed infestation and flower bud inoculation on systemic infection of turnip rape by *Albugo candida*. Can J Plant Sci 60:267–271

Verma U, Bhowmik TP (1989) Inheritance of resistance to a *Brassica juncea* pathotype of *Albugo candida* in *Brassica napus*. Can J Plant Pathol 11:443–444

Vicente JG, Conway J, Roberts SJ, Taylor JD (2001) Identification and origin of *Xanthomonas campestris* pv. *campestris* races and related pathovars. Phytopathology 91:492–499

Vicente JG, Gunn N, Bailey L, Pink D, Holub EB (2012) Genetics of resistance to downy mildew in *Brassica oleracea* and breeding towards durable disease control for UK vegetable production. Plant Pathol 61:600–609

Vicente JG, Holub EB (2013) *Xanthomonas campestris* pv. *campestris* (cause of black rot of crucifers) in the genomic era is still a worldwide threat to *Brassica* crops. Mol Plant Pathol 14:2–18

Voorrips RE, Jongerius MC, Kanne HJ (1997) Mapping of two genes for resistance to clubroot (*Plasmodiophora brassicae*) in a population of doubled haploid lines of *Brassica oleracea* by means of RFLP and AFLP markers. Theor Appl Genet 94:75–82

Wang Y, Lu W, Li J, Jiang Y (2006) Differential expression of two expansion genes in developing fruit of cracking-susceptible and -resistant litchi cultivars. J Amer Soc Hort Sci 131:118–121

Wang GX, Tang Y, Yan H, Sheng XG, Hao WW, Zhang L, Lu K, Liu F (2011) Production and characterization of interspecific somatic hybrids between *Brassica oleracea* var. *botrytis* and *B. nigra* and their progenies for the selection of advanced pre-breeding materials. Plant Cell Rep 30:1811–1821

Wang J, Gu H, Yu H, Zhao Z, Sheng X, Zhang X (2012) Genotypic variation of glucosinolates in broccoli (*Brassica oleracea* var. *italica*) florets from China. Food Chem 133:735–741

Wang Z, Fang H, Chen Y, Chen K, Li G, Gu S, Tan X (2014) Overexpression of *BnWRKY33* in oilseed rape enhances resistance to *Sclerotinia sclerotiorum*. Mol Plant Pathol 15:677–689

Wang GX, Lv J, Zhang J, Han S, Zong M, Guo N, Zeng XY, Zhang YY, Wang YP, Liu F (2016) Genetic and epigenetic alterations of *Brassica nigra* introgression lines from somatic hybridization: a resource for cauliflower improvement. Front Plant Sci 7:1258

Wei DY, Mei J, Fu Y, Disi JO, Li J, Qian W (2014) Quantitative trait loci analyses for resistance to *Sclerotinia sclerotiorum* and flowering time in *Brassica napus*. Mol Breed 34:1797–1804

Wei L, Jian H, Lu K, Filardo F, Yin N, Liu L, Qu C, Li W, Du H, Li J (2016) Genome-wide association analysis and differential expression analysis of resistance to *Sclerotinia* stem rot in *Brassica napus*. Plant Biotechnol J 14:1368–1380

Werner S, Diederichsen E, Frauen M, Schondelmaier J, Jung C (2007) Genetic mapping of clubroot resistance genes in oilseed rape. Theor Appl Genet 116:363–372

Westman AL, Kresovich S, Dickson MH (1999) Regional variation in *Brassica nigra* and other weedy crucifers for disease reaction to *Alternaria brassicicola* and *Xanthomonas campestris* pv. *campestris*. Euphytica 106:253–259

White PJ, Hammond JP, King GJ, Bowen HC, Hayden RM, Mecham MC, Spracklen WP, Broadley MR (2010) Genetic analysis of potassium use efficiency in *Brassica oleracea* L. Ann Bot 105:1199–1210

Williams PH (1985). White rust [*Albugo candida* (Pers. ex. Hook.) Kuntze]. In: Crucifer Genetics Cooperative (CRGC) Resource Book. University Wisconsin, USA, pp 1–7

Wu J, Cai G, Tu J, Li L, Liu S, Luo X, Zhou L, Fan C, Zhou Y (2013) Identification of QTLs for resistance to *Sclerotinia* stem rot and *BnaC.IGMTa* as a candidate gene of the major resistant QTL SRC6 in *Brassica napus*. PLoS One 8:e67740

Wu J, Zhao Q, Yang Q, Liu H, Li Q, Yi X, Cheng Y, Guo L, Fan C, Zhou Y (2016) Comparative transcriptomic analysis uncovers the complex genetic network for resistance to *Sclerotinia sclerotiorum* in *Brassica napus*. Sci Rep 6:19007

Xing M, Lv H, Ma J, Xu D, Li H, Yang L, Kang J, Wang X, Fang Z (2016) Transcriptome profiling of resistance to *Fusarium oxysporum* f. sp. *conglutinans* in cabbage (*Brassica oleracea*) roots. PLoS One11: e0148048

Yang B, Jiang Y, Rahman MH, Deyholos MK, Kav NNV (2009) Identification and expression analysis of *WRKY* transcription factor genes in canola (*Brassica napus* L.) in response to fungal pathogens and hormone treatments. BMC Plant Biol 9:68

Yi GE, Robin AH, Yang K, Park JI, Kang JG, Yang TJ, Nou IS (2015) Identification and expression analysis of glucosinolate biosynthetic genes and estimation of

glucosinolate contents in edible organs of *Brassica oleracea* subspecies. Molecules 20:13089–13111
Zhou Z, Weeden NF, Dickson MH (1997) The expression of a resistant gene to black rot in progeny of the protoplast fusion broccoli (*B. oleracea*). Cruciferae Newsl 19:109–110
Zhuang M, Zhang Y, Fang Z, Liu Y, Yang L, Sun P (2009) Studies on combining ability and heritability of splitting resistance characteristic in cabbage (*Brassica oleracea* var. *capitata* L.). China Veg 1:12–15

Genomic Research in *Brassica oleracea*: Retrospective and Future Prospects

11

Rod Snowdon, Shengyi Liu, and Chittaranjan Kole

Abstract

Brassica oleracea includes the economically most important C-genome representatives of U's famous *Brassica* triangle, spanning the vast range of European cabbage subspecies and varieties. Besides providing a basis for trait elucidation and genomics-assisted breeding, the *B. oleracea* genome also provides an important example and unique insight into ancient genome restructuring and post-polyploidization processes leading to patterning through genome fractionation. Looking forward, reference genome templates provide a basis for large-scale population genomics to capture the power of high-throughput sequencing technologies for application in crop improvement and improved biological understanding of important traits. The field of genomics is racing from one revolution to the next; however, and even before this book is published the latest DNA sequencing technologies have put scientists in a position to rapidly and cost-effectively generate ever better quality reference assemblies for complex crop genomes, including *B. oleracea.* Simultaneously, it is becoming increasingly feasible to translate knowledge about genes and genomes into extremely targeted crop improvement, using the latest gene-editing technologies. Future foci are, therefore, likely to encompass comprehensive pangenomic analyses of species-wide diversity, including more detailed knowledge about the relationships within and among major *Brassica* crop species and how this knowledge can be used for cross-species improvement, along with concerted efforts to translate population genomics into trait improvement by combining multi-level 'omics' approaches with biotechnological advances and classical breeding techniques. Genetic gain in breeding can benefit most greatly from acceleration of the time needed to advance generations or perform selection steps, and the *B. oleracea* genome lays a foundation for many different applications of genomics to improve efficiency and speed of key processes in this regard. This chapter outlines future prospects for implementation of genomics toward

R. Snowdon (✉)
Department of Plant Breeding, IFZ Research Centre for Biosystems, Land Use and Nutrition, Justus Liebig University, Heinrich-Buff-Ring 26-32, 35392 Giessen, Germany
e-mail: rod.snowdon@agrar.uni-giessen.de

S. Liu
Key Laboratory of Biology and Genetic Improvement of Oil Crops, The Ministry of Agriculture and Rural Affairs, Oil Crops Research Institute, Chinese Academy of Agricultural Sciences, Wuhan 430062, China

C. Kole
ICAR-National Institute for Plant Biotechnology, PUSA Campus, New Delhi 110012, India

S. Liu et al. (eds.), *The Brassica oleracea Genome*, Compendium of Plant Genomes,
https://doi.org/10.1007/978-3-030-31005-9_11

advancement of *B. oleracea* and related species via knowledge-based analysis and crop improvement.

11.1 Introduction

In the *Brassica* species triangle popularized by the Korean botanist WooJang-choon, better known by his Japanized name Nagaharu U (UN 1935), *Brassica oleracea* represents the diploid C-genome species which is not only itself a globally important crop but also contributed the C-subgenome to the important allopolyploid oilseed crops *B. carinata* (Ethiopian mustard) and *B. napus* (rapeseed, canola). As such, the genome of *B. oleracea* is representative not only for a globally important vegetable species complex but also for a vast range of closely related and inter-crossable C-genome Brassica species that serve as a highly valuable and enormously diverse germplasm resource for *Brassica* vegetable and oilseed improvement. In an age of unthinkable advances in genomic technologies and analytical power of large-scale DNA/RNA sequencing datasets, deep knowledge of plant genomes can provide a platform for significant advances in biological understanding of complex traits, discovery and generation of novel trait variation, and increasingly rapid, knowledge-based improvement of modern crop varieties. The prerequisite for a full and systematic exploitation of genomics technologies is the availability of reliable and well-annotated reference genome assemblies that can be used as a template for variant calling in order to associate genotypic differences with relevant biological characters and their underlying regulatory processes.

The first diploid *Brassica* genome assembly was that of *B. rapa* (Wang et al. 2011), chosen not only because of its own economic importance as a crop but also because it represented the smaller, less highly repetitive sub-genome of *B. napus*, a globally important oilseed crop. A few years later, as next-generation sequencing technologies evolved further and longer read lengths enabled more accurate and efficient assemblies even from more repetitive genomes, it became logical to also generate genome assemblies from *B. oleracea*, which represents the closely related A-subgenome in U's *Brassica* triangle. Two publications in 2014 reported genome assemble for two homozygous *B. oleracea* accessions, one a homozygous doubled-haploid line from the rapid-cycling accession TO1000 (Parkin et al. 2014) and the other a homozygous inbred cabbage -breeding line (Liu et al. 2014). Given the high quantity of repetitive sequences, it was inevitable that the shotgun-sequencing approach with predominantly short sequence reads could not completely cover the entire lengths of all chromosomes; however, a very high proportion of the gene spaces was captured and the genome sequences provided enormous advances in the ability to relate genetic information to annotated genes and regulatory information via integrated genetic mapping and genomic analyses.

Nevertheless, as ever-improving genomics technologies continue to challenge our imaginations, a genome reference will always be measured in relation to the next-best emerging technical advances. In the short time, some of the first *B. oleracea* genomes were published, long-read DNA sequencing methods and costs have advanced to a point where they are now becoming quite accessible even for researchers working with complex crop genomes. The most convincing argument for implementation of longer read lengths for assembly of complex, highly repetitive genomes like that of *B. oleracea* is the improvement in contiguity of scaffolds. Almost inevitably, before this book goes to press the first *B. oleracea* genome assembly has been published based on long-read sequencing using the Nanopore platform (Oxford Nanopore Technologies, Oxford, UK) in combination with optical mapping (Lam et al. 2012) on a Saphyr system from BioNanoGenomics (San Diego, CA, USA) and short-read data from an Illumina machine (Illumina, San Diego, CA, USA). Using the Nanopore minION system, a unique sequencing machine with negligible hardware costs, Belser et al. (2018) completed a new *B. oleracea* assembly from the broccoli cultivar HDEM. In this chapter, we explore how new genome sequencing technologies and new assembly methods are already expanding the scope of genome analysis in *B. oleracea*. These developments are discussed in the framework of potential impacts of present and future genome resources on *B. oleracea* genetics and breeding.

11.2 Advances in Genome Contiguity and Genome Structure-Function Analyses

The new *B. oleracea* assembly published recently by Belser et al. (2018), using Oxford Nanopore long-read data in combination with optimal mapping, represents a huge advance in contiguity in comparison to the previous assemblies based on short-read technologies. Importantly, almost 85% of the genome was anchored to the nine chromosomes, compared to around 70% in the previous assemblies. This demonstrates the increasing power of new sequencing and assembly technologies and underlines the importance of constant adjustments and updates of existing reference genomes with improved datasets. Belser et al. (2018) achieved a contig N50 of around 7.5 Mb, and the assembled HDEM genome could be represented by just 129 scaffolds, 20 of which resolve at almost chromosome arm level on the 9 *B. oleracea* chromosomes (Fig. 11.1).

Importantly, the long-read assembly resulted in anchoring of a total of 528.8 Mb, a considerable improvement on previously published assemblies, and helped considerably to resolve regions rich in transposable elements (Belser et al. 2018). For example, around 2000 more genes were included in the assembly than in the TO1000 assembly reported by Parkin et al. (2014). In coming years, it can be expected that average read lengths on Nanopore and other long-read sequencing platforms will continue to improve, and already the main obstacle to extremely long reads is the quality of high molecular weight genome DNA from plant samples. As the throughput increases and costs decrease (e.g., through implementation of the new Oxford NanporeTechnologies PromethIONmachines), extremely high-quality genome assemblies for B. oleracea are already possible for just a few thousand dollars and this price will continue to decrease. Thus, it will likely not take long until complete, chromosome-length assemblies will be available for multiple *B. oleracea* morphotypes as a basis for exact, high-resolution genomics.

High-quality assemblies are essential for complete elucidation of the role which transposable elements and other dispersed repeats play in genome evolution and regulation of gene/trait expression in complex (meso-) polyploid genomes. The availability of gold standard *B. oleracea* genomes will also provide a template to catalogue and describe genome-wide patterns of small RNA expression and the role of epigenetic factors in inheritance and trait expression. For example, one first study on genome-wide cytosine methylation identified methylation patterns in pollen development genes between a male-sterile cabbage and its fertile, near-isogenic maintainer line (Han et al. 2017), whereas differential miRNA expression was also found to associate with early pollen development in broccoli (Li et al. 2017).

11.3 Ancestral Genome Restructuring and *Brassica* Evolution

The most-extensive genome resequencing dataset published to date for *B. oleracea* (Cheng et al. 2016) spanned 119 accessions, representing leafy, heading, and tuberous vegetable morphotypes from Asia and Europe, and compared their genome sequences to a similarly broad range of genetic and morphotype diversity in 199 *B. rapa* accessions. Analyses of the sequence data revealed the major role of mesohexaploidization and subsequent genome fragmentation through mutation and/or loss of paralogous gene copies on morphotype diversification as a basis for natural and human selection. Interestingly, very similar conclusions can be drawn from a detailed analysis of resequencing data from *B. napus*, which underwent allopolyploidization just a few

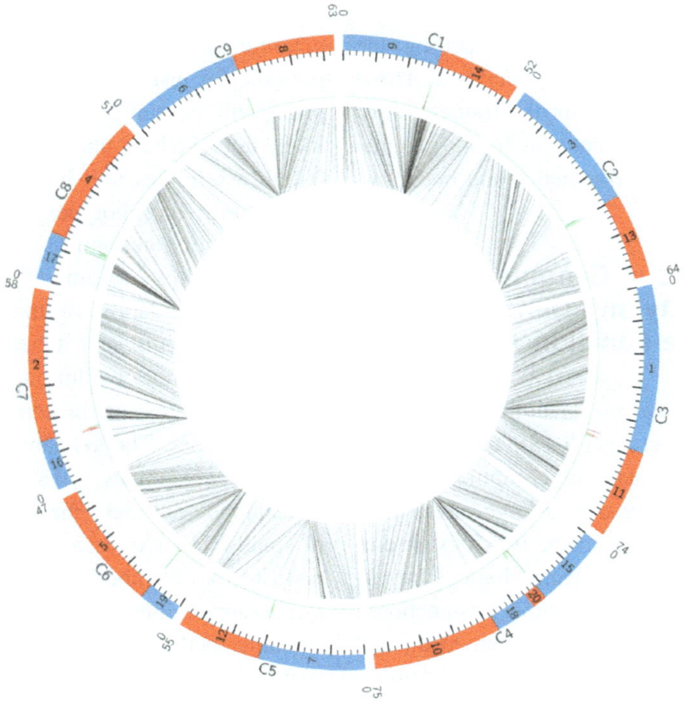

Fig. 11.1 Using Nanopore long-read sequencing in combination with optical mapping, Belser et al. (2018) generated a new genome assembly for *B. oleracea* broccoli variety HDEM which resolved the majority of the genome into just 20 scaffolds. In total, 528.8 Mb of the 630 Mb *B. oleracea* genome are anchored to the genetic map. On all pseudo-molecules except that representing chromosome C4, these 20 scaffolds are assembled to near chromosome-arm resolution, whereas the long arm of chromosome C4 is represented by three major scaffolds. This image shows the alignment of the HDEM scaffold assembly to the *B. oleracea* SNP linkage map from a cross between cultivars Richelain and HDEM. Red and blue bars around the outer perimeter represent assembled scaffolds, while the red and green peaks in the plot outside the genetic map show the positions of centromeres (*CenBr1* repeats in red and *CenBr2* repeats in green). Reprinted with permission from Belser et al. (2018) Chromosome-scale assemblies of plant genomes using nanopore long reads and optical maps. Nature Plants 2018, © Springer Nature

thousand years ago (Chalhoub et al. 2014). A number of recent studies have demonstrated that post-polyploidization genome restructuring and segmentation chromosome exchanges, caused by illegitimate intergenomic pairing of homoeologous chromosomes, appear to be a major and ongoing driver of genome evolution and selective variation in *B. napus*(Chalhoub et al. 2014; Hurgobin et al. 2017; Samans et al. 2017). A consequence of interhomoeolog exchanges appears to be accelerated rates of mutation and presence-absence variation, which in *B. napus* led to the creation of de novo variation for eco-geographical flowering and morphotype traits (Schiessl et al. 2017) as a basis for agricultural selection. This raises interesting parallels to the genome restructuring patterns associated with mesohexaploidizatonin *Brassica* diploids and the subsequent emergence and domestication of convergent morphotypes in *B. rapa* and *B: oleracea*.

A pangenome analysis spanning nine cultivated *B. oleracea* cultivars and a wild C-genome relative (*Brassica macrocarpa*) revealed a strong role of presence-absence variation in the evolution of *B. oleracea* crop forms (Golicz et al. 2016). Of around 61,000 genes, they identified in the 10 accessions, some 18% were reported to show presence-absence variation (PAV) among the 10 sequenced accessions. Genes affected by

PAV included major functions and/or pathways of agronomic importance in *B. oleracea*, for example, flowering time, disease resistance, glucosinolate metabolism, or vitamin biosynthesis. The ability in the foreseeable future to perform pangenomic analyses implementing fully assembled, high-quality genomes from large numbers of genetically diverse accessions will provide comprehensive insight into both large- and small-scale structural variations associated with agronomic selection. Already, long-read skim sequencing provides a reasonably inexpensive and extremely effective method to localzse PAV associating with important traits at a single-gene or even intragenic scale (Gabur et al. 2018b). Future analyses will reveal the extent of such variants and their trait associations in large collections of *B. oleracea* and related C-genome Brassica species. Indeed, the *B. oleracea* genome provides the basis for the first detailed analysis and sequence mining of wild C-genome species as a rice resource for genetic improvement in crop brassicas. Genetic diversity in the economically most important allotetraploid *Brassica* oilseed crop *B. napus*(rapeseed) is particularly low in the C-subgenome, and sequence diversity identified by mining of available diversity using reference genomes or the pangenome as a template can be a highly valuable source of many traits including disease resistance (Rygulla et al. 2007; Mei et al. 2015), seed quality (Jesske et al. 2013b), insect resistance (Jesske et al. 2013b) or heterosis (Jesske et al. 2013a; Snowdon et al. 2015), for example.

As evidence grows that PAV associates with complex quantitative traits in recent *Brassica* allopolyploids (He et al. 2017; Gabur et al. 2018a), there is good reason to believe that closer investigations of trait-sequence associations in the mesopolyploid *Brassica* species will reveal the remnants of equally important selective and adaptive variation for important traits in *B. oleracea*. High-quality reference sequences provide a highly valuable platform to achieve this goal by improving resolution of paralogous gene copies and revealing functional sequence variation associating with important traits. For example, by reanalyzing the resequencing data from the 119 *B. oleracea* accessions described by Cheng et al. (2016) using the HDEM reference genome, Belser et al. (2018) revealed a considerable proportion of previously non-captured coding sequences. An increased level of detail and accuracy in discrimination between paralogues of complex gene families, for example, the strongly duplicated *FLOWERING LOCUS C* (*FLC*) genes, gives deep insight into presence-absence and/or functional sequence variants discriminating different crop morphotypes or other essential traits (Belser et al. 2018; Snowdon and Schiessl 2019; Wu et al. 2019).

11.4 The Growing Need for an Integrated *Brassica* Bioinformatics Platform

Beyond the crop brassicas, it is now also feasible to generate high-quality reference sequences for crop wild relatives as well as pangenomes within the U triangle. In association with ever-growing data accumulation from resequencing, transcriptomics, and other analyses with large germplasm collations and experimental populations, these will represent a hugely valuable data resource for germplasm management and future improvement of *B. oleracea* crops (and its close relatives) by genomics-assisted breeding. To achieve this goal, however, there is an urgently growing need for coordinated, integrated platforms to make genomics data from large populations available in a useful manner for the international research community. A key aspect which needs urgent consideration is the adoption and implementation of standardized and flexible annotation systems for rapidly growing quantities and depths of gene and (pan-)genome datasets. Ideally, a system which allows cross-species comparisons of gene names and positions, sequence variants, and functional annotations among the species of the Brassica triangle would be a highly valuable starting point, while ultimately links to genome information and annotations for Arabidopsis and other Brassicaceae would be highly desirable in order to establish and examine syntenic and structural inter-genome relationships and investigate the genome dynamics of cross-genera (crop) evolution.

A number of public Brassica genomics databases are already available for searches and comparative analyses, and up-to-date hyperlinks can be found via the community portal www.brassica.info. A major challenge facing not just Brassica researchers is the development and maintenance of platforms for analysis and visualization of genome data in the context of high-density genetic maps and complex trait/QTL data for public germplasm collections. Future databases must seek workable solutions for integration of genetic materials, genomics, transcriptomics, metabolomics, proteomics, and phenomics, implementing search and comparison tools with visual outcomes. As genetics and breeding move more and more heavily toward an implementation of and reliance on big data and artificial intelligence, but where end-users are simultaneously becoming less and less connected to the data they wish to process and use, the need for effective informatics tools to translate genomics into solutions for practical applications is becoming an increasingly important goal. These are common problems which all *Brassica* researchers need to address collectively across species, at the same time also observing and learning from developments for other crop and model plants.

11.5 Genomics-Assisted Breeding

During the past 3-4 decades, marker-assisted selection became the method of choice for selecting simple, monogenic traits with difficult phenotypes in plant breeding programs. However, the widely held expectation after the invention of PCR-based marker technologies in association with detailed quantitative locus (QTL) analyses, namely, that genetic markers linked to QTL for important traits would also lead to huge advances in selection of more complex traits, has never been realized to a really satisfactory extent (Bernardo 2016). To a certain extent, the use of genome data, high-throughput genotyping techniques, genomic selection, and genome-based hybrid performance prediction to identify useful diversity and implement it in breeding programs has also not yet provided the breeding breakthroughs which might be expected from such powerful technologies. In contrast to PCR-based markers, however, genome-wide single nucleotide polymorphism (SNP) markers or other genomics technologies providing ultra-high genome resolution have a considerable advantage in that they come considerably closer to representing the infinitesimal model of quantitative trait inheritance. As such, despite similar pronouncements in the past, there seem to be more solid grounds to believe that predictions of breeding advances from genome-wide data will indeed lead to a revolution in breeding. Examples from other crops and particularly from animal breeding suggest that the availability of high-quality genome assemblies, vast genomic datasets, and large-scale digital phenomic datasets are set to play a key role in the future implementation of omics-assisted breeding in *B. oleracea* breeding.

Of all the components contributing to genetic gain in a breeding program, the most significant singular improvements can be achieved by acceleration of the time needed to make selection decisions or to advance generations. As such, crops like *B. oleracea* with long generation intervals—for example, due to a strong vernalization requirement and/or late-season flowering—stand to benefit particularly strongly from advances which accelerate breeding processes like generation advancement and/or selection. Hence, marker-assisted selection even for a simple Mendelian trait can provide an advantage over phenotypic selection if the trait is only visible in later stages of the plant life-cycle. Similarly, markers can reduce costly and time-consuming self-pollination and validation steps that are needed in a backcrossing program to identify heterozygous carriers of a desirable, recessive allele solely by phenotypic selection. Modern genome screening tools, for example, the 60K Illumina Brassica SNP-genotyping array (Clarke et al. 2016; Mason et al. 2017), provide dense genome-wide profiles that enable the capture of linkage-disequilibrium to more or less all genes and/or QTL in a genome. Ordering of high-density, genome-wide markers makes it relatively simple in suitable mapping populations to identify markers with tight linkage to major-effect or Mendelian loci controlling traits of agronomic interest. Hence, the genome sequence greatly facilitates the implementation of markers in marker-assisted breeding programs. Furthermore, fine-mapping or mapping-by-sequencing approaches in association with the genome sequence provide a highly powerful method to identify causal genes and functional mutations, and in principle the causal mutation itself can be implemented as the 'perfect marker' for selection of a trait of interest. Of particular interest to *B. oleracea* breeders are markers within or tightly linked to genes inferring leaf traits, for example, glossy (Liu et al. 2017b) or pigmented leaves (Li et al. 2016; Liu et al. 2017a), head traits like head-splitting susceptibility (Su et al. 2015), important disease resistances (Rubel et al. 2017; Shimizu et al. 2015) or male-sterility (Liang et al. 2017). As for other crops, the identification of functional mutations and cloning of genes associated with major traits is considerably facilitated by the availability of the genome sequence, and as resequencing costs continue to decrease it is becoming increasingly simple to identify function mutants and markers for breeding. A recent study in a cross between cabbage and broccoli demonstrated the enormous power of QTL bulk-sequencing (QTL-seq) to rapidly identify genes and causal mutations associated with monogenic or oligogeneic traits in *B. oleracea* using only a reference genome without relying on genetic linkage mapping (Shu et al. 2018). Using QTL-seq in an F2 mapping population (Fig. 11.2), the authors delineated a major early-flowering QTL on chromosome 2 to a 228 kb genomic region containing 29 genes. One of these, Bol024659, was found to be a homolog of a major gene regulating flowering in Arabidopsis and showed expression patterns indicating that it is a strong candidate for early-flowering in *B. oleracea* crops (Shu et al. 2018). Even for loci which stretch over multiple kilobases, for example, the S-locus in *B. oleracea*, resolution of allelic variants and structural variants will soon become routine using long-read sequence data that are able to provide contiguous sequences from multiple genotypes (Belser et al. 2018). It may only be a matter of a few years cost-effective QTLseq approaches are able to reliably identify causal genes even in complex genome regions like pericentromeric repeats.

As with other crops, advances in use of genetic markers for simple selection of complex traits is less well-advanced than for monogenic traits due to the poor transferability of highly-quantitative traits into different genetic backgrounds, or the low effect of many QTL in comparison to the environmental variance. On the other hand, genomic selection procedures have proven very effective for prediction of quantitative trait performance in animals and many other crops (Heffner et al. 2009; Lorenz et al. 2011) and when suitable gentling tools like the Brassica 60 k SNP array are available then such methods have exceptional promise to also advance breeding progress in *B. oleracea*. As long as accurate phenotype data can be gathered from large, representative training populations, it is feasible to more or less accurately estimate genetic variance contributions of genome-wide markers or haplotypes and complement these into genomic selection models. For *B. oleracea*, where hybrid breeding plays a key role but generation and testing of test-hybrids remains technically challenging and time-consuming, the implementation of genomic prediction of hybrid performance can provide a considerable advantage in terms of pre-selecting best predicted hybrid combinations. Published examples are rare in cabbage breeding so far, but these methods are already in use in commercial breeding operations and examples from commercial *B. napus* breeding populations show their robustness and power in accelerating breeding gain in *Brassica* crops (Werner et al. 2018).

11.6 Genome Editing

Chemical mutagenesis was occasionally used in the past for non-targeted modification of *B. oleracea* genes, and in isolated cases, this

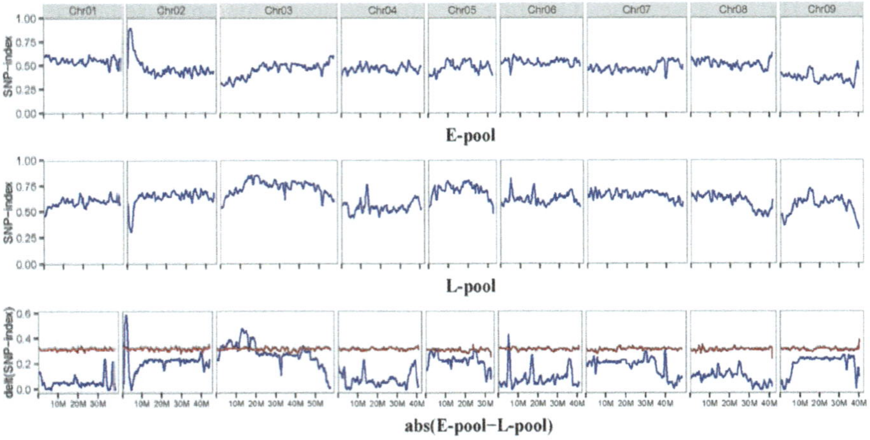

Fig. 11.2 Example for QTL-seq for rapid identification of candidate genes for quantitative traits based on bulked-segregant resequencing and read mapping to a *B. oleracea* reference sequence. Shu et al. (2018) sequenced F2 offspring from a cabbage x broccoli cross segregating for early (E-pool) and late (L-pool) flowering. Differential analysis of genome-wide SNP allele frequencies using the QTL-seq approach delineated a major QTL for early flowering on chromosome 2 containing a total of 29 genes, one of which was a strong candidate for flowering-time based on functional comparisons with Arabidopsis and expression studies. Reprinted with permission from Shu et al. (2018), QTL-seq for rapid identification of candidate genes for flowering time in broccoli × cabbage. Theoretical and Applied Genetics 2018, © Springer Nature

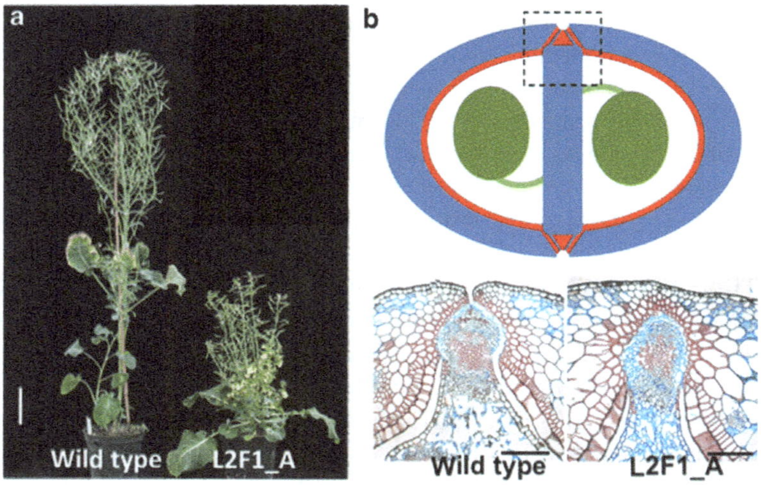

Fig. 11.3 The first report describing the use of CRISPR-Cas9 genome editing for targeted gene modification in *B. oleracea* was published by Lawrensen et al. (2015). Induced mutations in the gibberellin acid synthesis gene *BolC.GA4.a* resulted in dwarf stature of plants and affect the valve margin of siliques in mutated plants. **a** Wild-type *B. oleracea* DH1012 (left) in comparison to mutant L2F1_A, which shows a severe dwarf phenotype due to altered gibberellin status. Scale bar 10 cm. **b** Schematic cross section of a normal *B. oleracea* silique (above) along with microscopic cross sections from siliques of wild-type and L2F1_A plants (below). The valve margin region is indicated by dashed square in the cartoon. Lignified tissue is indicated in red, unlignified cells in blue, while developing seeds in the carton are shown in green. Scale bars: 200 μm. Image reproduced with slight modifications from Lawrensen et al. (2015); https://doi.org/10.1186/s13059-015-0826-7) under the terms of the Creative Commons Attribution licence 2.0. © Lawrenson et al. (2015)

approach has enabled direct discovery of potentially useful mutations when combined with targeted gene sequencing. One example is the identification of mutagenized *B. oleracea* seeds for generation of mutants with altered fatty acid dehydrogenase genes (Rahman et al. 2013). However, such approaches result in extremely high levels of off-target mutations which require considerable downstream backcrossing effort for effective use in breeding. In contrast, the discovery of highly efficient and accurate techniques for targeted gene editing (Cong et al. 2013; Shan et al. 2013) promises a revolution in crop genetics, with an unprecedented potential to generate novel, targeted mutations in any known gene.

Although still in its infancy and limited to crops and genotypes amenable to tissue culture regeneration, first applications of genome editing via clustered regularly interspaced short palindromic repeats (CRISPR) in association with CRISPR-associated protein 9 (Cas9) have shown that gene editing can also be implemented to generate targeted mutations in *B. oleracea*. In the first example, targeting of the *B. oleracea* gibberellin biosynthesis gene *BolC.GA4.a* resulted in a 10% induction rate of Cas9-induced mutations among first-generation plants that resulted in T0 plants with the expected dwarf phenotype caused by a *GA4* knockout (Lawrenson et al. 2015) (Fig. 11.3). This study also demonstrated the ability to specifically target multiple copies of the same gene in polyploid Brassica, a potential advantage of CRISPR-Cas9 in comparison to previously available mutagenesis techniques. In another, more recent example, orthologous copies of the carotenoid biosynthesis gene *PHYTOENE DESATURASE* (*BolPDS*) and the vernalization gene *FRIGIDA* (*BolFRI*) gene were targeted in cabbage with DNA-free editing constructs designed using guide RNAs and primers conserved across *Brassica* species (Murovec et al. 2018). The ability for DNA-free editing provides a potential possibility to overcome regulatory constraints associated with non-target vector insertions, although regulation of *cis*-genic CRISPR gene editing in crop plants is already very different in different regions, and particularly Europe must first establish a suitable regulatory framework capable of dealing with the classification of CRISPR-modified plants as genetically modified organisms (GMO) even though they may not be distinguishable from other kinds of non-GMO natural or induced mutants (Eriksson 2018).

References

Belser C, Istace B, Denis E, Dubarry M, Baurens FC, Falentin C, Genete M, Berrabah W, Chevre AM, Delourme R, Deniot G, Denoeud F, Duffe P, Engelen S, Lemainque A, Manzanares-Dauleux M, Martin G, Morice J, Noel B, Vekemans X, D'Hont A, Rousseau-Gueutin M, Barbe V, Cruaud C, Wincker P, Aury JM (2018) Chromosome-scale assemblies of plant genomes using nanopore long reads and optical maps. Nat Plants 4:879–887

Bernardo R (2016) Bandwagons I, too, have known. Theor Appl Genet 129:2323–2332

Chalhoub B, Denoeud F, Liu SY, Parkin IAP, Tang HB, Wang XY, Chiquet J, Belcram H, Tong CB, Samans B, Correa M, Da Silva C, Just J, Falentin C, Koh CS, Le Clainche I, Bernard M, Bento P, Noel B, Labadie K, Alberti A, Charles M, Arnaud D, Guo H, Daviaud C, Alamery S, Jabbari K, Zhao MX, Edger PP, Chelaifa H, Tack D, Lassalle G, Mestiri I, Schnel N, Le Paslier MC, Fan GY, Renault V, Bayer PE, Golicz AA, Manoli S, Lee TH, Thi VHD, Chalabi S, Hu Q, Fan CC, Tollenaere R, Lu YH, Battail C, Shen JX, Sidebottom CHD, Wang XF, Canaguier A, Chauveau A, Berard A, Deniot G, Guan M, Liu ZS, Sun FM, Lim YP, Lyons E, Town CD, Bancroft I, Wang XW, Meng JL, Ma JX, Pires JC, King GJ, Brunel D, Delourme R, Renard M, Aury JM, Adams KL, Batley J, Snowdon RJ, Tost J, Edwards D, Zhou YM, Hua W, Sharpe AG, Paterson AH, Guan CY, Wincker P (2014) Earlyallopolyploidevolution in the post-Neolithic Brassicanapu soil seed genome. Science 345:950–953

Cheng F, Sun RF, Hou XL, Zheng HK, Zhang FL, Zhang YY, Liu B, Liang JL, Zhuang M, Liu YX, Liu DY, Wang XB, Li PX, Liu YM, Lin K, Bucher J, Zhang NW, Wang Y, Wang H, Deng J, Liao YC, Wei KY, Zhang XM, Fu LX, Hu YY, Liu JS, Cai CC, Zhang SJ, Zhang SF, Li F, Zhang H, Zhang JF, Guo N, Liu ZY, Liu J, Sun C, Ma Y, Zhang HJ, Cui Y, Freeling MR, Borm T, Bonnema G, Wu J, Wang XW (2016) Subgenomeparallelselectionisassociatedwith morphotype diversification and convergent crop domestication in Brassicarapa and Brassicaoleracea. Nature Genet 48:1218–1224

Clarke WE, Higgins EE, Plieske J, Wieseke R, Sidebottom C, Khedikar Y, Batley J, Edwards D, Meng J, Li R, Lawley CT, Pauquet J, Laga B, Cheung W,

Iniguez-Luy F, Dyrszka E, Rae S, Stich B, Snowdon RJ, Sharpe AG, Ganal MW, Parkin IA (2016) A high-density SNP genotyping array for Brassica napus and its ancestral diploid species based on optimised selection of single-locus markers in the allotetraploid genome. Theor Appl Genet 129:1887–1899

Cong L, Ran FA, Cox D, Lin S, Barretto R, Habib N, Hsu PD, Wu X, Jiang W, Marraffini LA, Zhang F (2013) Multiplex genome engineering using CRISPR/Cas systems. Science 339:819–823

Eriksson D (2018) The evolving EU regulatory framework for precision breeding. TheorAppl Genet 132:569–573

Gabur I, Chawla HS, Liu XW, Kumar V, Faure S, von Tiedemann A, Jestin C, Dryzska E, Volkmann S, Breuer F, Delourme R, Snowdon R, Obermeier C (2018a) Finding invisible quantitative trait loci with missing data. Plant Biotechnol J 16:2102–2112

Gabur I, Chawla HS, Snowdon RJ, Parkin IAP (2018b) Connecting genome structural variation with complex traits in crop plants. Theor Appl Genet 132:733–750

Golicz AA, Bayer PE, Barker GC, Edger PP, Kim H, Martinez PA, Chan CKK, Severn-Ellis A, McCombie WR, Parkin IAP, Paterson AH, Pires JC, Sharpe AG, Tang HB, Teakle GR, Town CD, Batley J, Edwards D (2016) The pangenome of an agronomically important crop plant *Brassicaoleracea*. NatCommun 7:13390

Han F, Zhang X, Liu X, Su H, Kong C, Fang Z, Yang L, Zhuang M, Zhang Y, Liu Y, Li Z, Lv H (2017) Comparative analysis of genome wide DNA methylation profiles for the genic male sterile cabbage line 01-20S and its maintainer line. Genes (Basel) 8(6):159

He Z, Wang L, Harper AL, Havlickova L, Pradhan AK, Parkin IAP, Bancroft I (2017) Extensive homoeologous genome exchanges in allopolyploid crops revealed by mRNAseq-based visualization. Plant Biotechnol J 15:594–604

Heffner EL, Sorrells ME, Jannink JL (2009) Genomic Selection for Crop Improvement. Crop Sci 49:1–12

Hurgobin B, Golicz AA, Bayer PE, Chan CK, Tirnaz S, Dolatabadian A, Schiessl SV, Samans B, Montenegro JD, Parkin IAP, Pires JC, Chalhoub B, King GJ, Snowdon R, Batley J, Edwards D (2017) Homoeologous exchange is a major cause of gene presence/absence variation in the amphidiploid *Brassica napus*. Plant Biotechnol J16:1265–1274

Jesske T, Olberg B, Schierholt A, Becker HC (2013a) Resynthesized lines from domesticated and wild Brassica taxa and their hybrids with *Bnapus* L.: genetic diversity and hybrid yield. Theor Appl Genet 126:1053–1065

Jesske T, Schierholt A, Becker HC (2013b) Glucosinolates and other seed quality traits of resynthesized *Brassica napus* L. derived from domesticated and wild Brassica taxa. Genet Resour Crop Evol 60:2273–2282

Lam ET, Hastie A, Lin C, Ehrlich D, Das SK, Austin MD, Deshpande P, Cao H, Nagarajan N, Xiao M, Kwok PY (2012) Genome mapping on nanochannel arrays for structural variation analysis and sequence assembly. Nat Biotechnol 30:771–776

Lawrenson T, Shorinola O, Stacey N, Li C, Østergaard L, Patron N, Uauy C, Harwood W (2015) Induction of targeted, heritable mutations in barley and Brassica oleracea using RNA-guided Cas9 nuclease. Genome Biol 16

Li H, Wang Y, Wu M, Li L, Jin C, Zhang Q, Chen C, Song W, Wang C (2017) Small RNA sequencing reveals differential miRNA expression in the early development of broccoli (*Brassica oleracea* var. *italica*) pollen. Front Plant Sci 8:404

Li H, Zhu L, Yuan G, Heng S, Yi B, Ma C, Shen J, Tu J, Fu T, Wen J (2016) Fine mapping and candidate gene analysis of an anthocyanin-rich gene, *BnaA.PL1*, conferring purple leaves in *Brassica napus* L. Mol Genet Genom 291:1523–1534

Liang J, Ma Y, Wu J, Cheng F, Liu B, Wang X (2017) Map-based cloning of the dominant genic male sterile *Ms-cd1* gene in cabbage (*Brassica oleracea*). Theor Appl Genet 130:71–79

Liu S, Liu Y, Yang X, Tong C, Edwards D, Parkin IA, Zhao M, Ma J, Yu J, Huang S, Wang X, Wang J, Lu K, Fang Z, Bancroft I, Yang TJ, Hu Q, Wang X, Yue Z, Li H, Yang L, Wu J, Zhou Q, Wang W, King GJ, Pires JC, Lu C, Wu Z, Sampath P, Wang Z, Guo H, Pan S, Yang L, Min J, Zhang D, Jin D, Li W, Belcram H, Tu J, Guan M, Qi C, Du D, Li J, Jiang L, Batley J, Sharpe AG, Park BS, Ruperao P, Cheng F, Waminal NE, Huang Y, Dong C, Wang L, Li J, Hu Z, Zhuang M, Huang Y, Huang J, Shi J, Mei D, Liu J, Lee TH, Wang J, Jin H, Li Z, Li X, Zhang J, Xiao L, Zhou Y, Liu Z, Liu X, Qin R, Tang X, Liu W, Wang Y, Zhang Y, Lee J, Kim HH, Denoeud F, Xu X, Liang X, Hua W, Wang X, Wang J, Chalhoub B, Paterson AH (2014) The Brassica oleracea genome reveals the asymmetrical evolution of polyploid genomes. Nat Commun 5:3930

Liu XP, Gao BZ, Han FQ, Fang ZY, Yang LM, Zhuang M, Lv HH, Liu YM, Li ZS, Cai CC, Yu HL, Li ZY, Zhang YY (2017a) Genetics and fine mapping of a purple leaf gene, BoPr, in ornamental kale (*Brassica oleracea* L. var. *acephala*). BMC Genomics 18:230

Liu Z, Fang Z, Zhuang M, Zhang Y, Lv H, Liu Y, Li Z, Sun P, Tang J, Liu D, Zhang Z, Yang L (2017b) Fine-mapping and analysis of *Cgl1*, a gene conferring glossy trait in cabbage (*Brassicaoleracea* L. var. *capitata*). Front Plant Sci 8:239

Lorenz AJ, Chao SM, Asoro FG, Heffner EL, Hayashi T, Iwata H, Smith KP, Sorrells ME, Jannink JL (2011) Genomic selection in plant Breeding: Knowledge and prospects. Adv Agron 110:77–123

Mason AS, Higgins EE, Snowdon RJ, Batley J, Stein A, Werner C, Parkin IA (2017) A user guide to the Brassica 60 K Illumina Infinium SNP genotyping array. Theor Appl Genet 130:621–633

Mei JQ, Liu Y, Wei DY, Wittkop B, Ding YJ, Li QF, Li JN, Wan HF, Li ZY, Ge XH, Frauen M,

Snowdon RJ, Qian W, Friedt W (2015) Transfer of sclerotinia resistance from wild relative of Brassica oleracea into Brassica napus using a hexaploidy step. TheorAppl Genet 128:639–644

Murovec J, Gucek K, Bohanec B, Avbelj M, Jerala R (2018) DNA-free genome editing of *Brassica oleracea* and *B. rapa*protoplasts using CRISPR-Cas9 ribonucleoprotein complexes. Front Plant Sci 9:1594

Parkin IA, Koh C, Tang H, Robinson SJ, Kagale S, Clarke WE, Town CD, Nixon J, Krishnakumar V, Bidwell SL, Denoeud F, Belcram H, Links MG, Just J, Clarke C, Bender T, Huebert T, Mason AS, Pires JC, Barker G, Moore J, Walley PG, Manoli S, Batley J, Edwards D, Nelson MN, Wang X, Paterson AH, King G, Bancroft I, Chalhoub B, Sharpe AG (2014) Transcriptome and methylome profiling reveals relics of genome dominance in the mesopolyploid Brassica oleracea. Genome Biol 15:R77

Rahman H, Singer SD, Weselake RJ (2013) Development of low-linolenic acid *Brassica oleracea* lines through seed mutagenesis and molecular characterization of mutants. Theor Appl Genet 126:1587–1598

Rubel MH, Robin AHK, Natarajan S, Vicente JG, Kim HT, Park JI, Nou IS (2017) Whole-genome re-alignment facilitates development of specific molecular markers for races 1 and 4 of *Xanthomonas campestris* pv. *campestris*, the cause of black rot disease in *Brassica oleracea*. Intl J Mol Sci 18(12): E2523

Rygulla W, Friedt W, Seyis F, Luhs W, Eynck C, von Tiedemann A, Snowdon RJ (2007) Combination of resistance to Verticillium longisporum from zero erucic acid Brassica oleracea and oilseed Brassica rapa genotypes in resynthesized rapeseed (*Brassica napus*) lines. Plant Breed 126:596–602

Samans B, Chalhoub B, Snowdon RJ (2017) Surviving a genome collision: Genomic signatures of allopolyploidization in the recent crop species *Brassica napus*. Plant Genome 10(3). https://doi.org/10.3835/plantgenome2017.02.0013

Schiessl S, Huettel B, Kuehn D, Reinhardt R, Snowdon R (2017) Post-polyploidisation morphotype diversification associates with gene copy number variation. Sci Rep 7:41845

Shan Q, Wang Y, Li J, Zhang Y, Chen K, Liang Z, Zhang K, Liu J, Xi JJ, Qiu JL, Gao C (2013) Targeted genome modification of crop plants using a CRISPR-Cas system. Nat Biotechnol 31:686–688

Shimizu M, Pu ZJ, Kawanabe T, Kitashiba H, Matsumoto S, Ebe Y, Sano M, Funaki T, Fukai E, Fujimoto R, Okazaki K (2015) Map-based cloning of a candidate gene conferring Fusarium yellows resistance in *Brassica oleracea*. Theor Appl Genet 128:119–130

Shu J, Liu Y, Zhang L, Li Z, Fang Z, Yang L, Zhuang M, Zhang Y, Lv H (2018) QTL-seq for rapid identification of candidate genes for flowering time in broccoli x cabbage. Theor Appl Genet 131:917–928

Snowdon RJ, Abbadi A, Kox T, Schmutzer T, Leckband G (2015) Heterotic Haplotype Capture: precision breeding for hybrid performance. Trends Plant Sci 20:410–413

Snowdon RJ, Schiessl SV (2019) Illuminating crop adaptation using population genomics. Mol Plant 12:27–29

Su Y, Liu Y, Li Z, Fang Z, Yang L, Zhuang M, Zhang Y (2015) QTL Analysis of head splitting resistance in cabbage (*Brassica oleracea* L. var. *capitata*) using SSR and InDel makers based on whole-genome re-sequencing. PLoS One 10:e0138073

UN (1935) Genome analysis in Brassica with special reference to the experimental formation of B. napus and peculiar mode of fertilization. Jpn J Bot 7:389–453

Wang X, Wang H, Wang J, Sun R, Wu J, Liu S, Bai Y, Mun JH, Bancroft I, Cheng F, Huang S, Li X, Hua W, Wang J, Wang X, Freeling M, Pires JC, Paterson AH, Chalhoub B, Wang B, Hayward A, Sharpe AG, Park BS, Weisshaar B, Liu B, Li B, Liu B, Tong C, Song C, Duran C, Peng C, Geng C, Koh C, Lin C, Edwards D, Mu D, Shen D, Soumpourou E, Li F, Fraser F, Conant G, Lassalle G, King GJ, Bonnema G, Tang H, Wang H, Belcram H, Zhou H, Hirakawa H, Abe H, Guo H, Wang H, Jin H, Parkin IA, Batley J, Kim JS, Just J, Li J, Xu J, Deng J, Kim JA, Li J, Yu J, Meng J, Wang J, Min J, Poulain J, Wang J, Hatakeyama K, Wu K, Wang L, Fang L, Trick M, Links MG, Zhao M, Jin M, Ramchiary N, Drou N, Berkman PJ, Cai Q, Huang Q, Li R, Tabata S, Cheng S, Zhang S, Zhang S, Huang S, Sato S, Sun S, Kwon SJ, Choi SR, Lee TH, Fan W, Zhao X, Tan X, Xu X, Wang Y, Qiu Y, Yin Y, Li Y, Du Y, Liao Y, Lim Y, Narusaka Y, Wang Y, Wang Z, Li Z, Wang Z, Xiong Z, Zhang Z, Brassica rapa Genome Sequencing Project C (2011) The genome of the mesopolyploid crop species *Brassica rapa*. Nat Genet 43:1035–1039

Werner CR, Qian LW, Voss-Fels KP, Abbadi A, Leckband G, Frisch M, Snowdon RJ (2018) Genome-wide regression models considering general and specific combining ability predict hybrid performance in oilseed rape with similar accuracy regardless of trait architecture. Theor Appl Genet 131:299–317

Wu D, Liang Z, Yan T, Xu Y, Xuan L, Tang J, Zhou G, Lohwasser U, Hua S, Wang H, Chen X, Wang Q, Zhu L, Maodzeka A, Hussain N, Li Z, Li X, Shamsi IH, Jilani G, Wu L, Zheng H, Zhang G, Chalhoub B, Shen L, Yu H, Jiang L (2019) Whole-genomeresequencing of a worldwidecollection of rapeseedaccessions reveals the genetic basis of ecotype divergence. Mol Plant 12:30–43

The manufacturer's authorised representative in the EU is Springer
Nature Customer Service Centre GmbH, Europaplatz 3, 69115 Heidelberg,
Germany. If you have any concerns regarding our products, please
contact ProductSafety@springernature.com

Printed and bound by CPI Group (UK) Ltd, Croydon, CR0 4YY
30/09/2026
02225665-0019